T0139920

Intelligent Integrated Energy Systems

Peter Palensky · Miloš Cvetković
Tamás Keviczky
Editors

Intelligent Integrated Energy Systems

The PowerWeb Program at TU Delft

 Springer

Editors
Peter Palensky
Electrical Sustainable Energy (ESE)
Delft University of Technology
Delft, Zuid-Holland, The Netherlands

Tamás Keviczky
Delft Center for Systems and Control
 (DCSC)
Delft University of Technology
Delft, Zuid-Holland, The Netherlands

Miloš Cvetković
Electrical Sustainable Energy (ESE)
Delft University of Technology
Delft, Zuid-Holland, The Netherlands

ISBN 978-3-030-13081-7   ISBN 978-3-030-00057-8   (eBook)
https://doi.org/10.1007/978-3-030-00057-8

This Springer imprint is published by the registered company Springer Nature Switzerland AG
The registered company address is: Gewerbestrasse 11, 6330 Cham, Switzerland

Preface

PowerWeb is TU Delft's consortium for interdisciplinary research on intelligent, integrated energy systems. In operation since 2012, it acts as a host and information platform for a growing number of projects and activities, ranging from single Ph.D. student projects up to large integrated and international research programs. PowerWeb acts in an interfaculty fashion and combines experts from electrical engineering, computer science, mathematics, mechanical engineering, technology and policy management, control engineering, civil engineering, architecture, aerospace engineering, and industrial design. Many activities are organized with and for students, such as lunch lectures of invited industrial speakers, or the annual PowerWeb conference which features a large Ph.D. project poster session.

Within PowerWeb, project questions are typically associated to or sorted into one of three problem domains:

- Grid technology: hardware, energy system, generation, transmission, distribution, use, conversion, pipes, storage, and more,
- Intelligence: for instance ICT, algorithms, data, operations, controls, management, balancing, security and quality of supply, analytics, and planning,
- Society: socioeconomic mechanisms, users, environmental goals, institutions, policies, markets, and so on.

PowerWeb is not limited to electricity. It bridges heat, gas, and other types of energy with markets, industrial processes, transport, and the built environment.

PowerWeb serves as a singular entry point for industry to the University's knowledge. Via its Industry Advisory Board, a steady link to business owners, manufacturers, and energy system operators is provided.

This book gives an overview on the theory of intelligent, integrated energy systems, and reports on several PowerWeb projects and their role in markets and institutions.

Part I covers a number of projects that were executed under the PowerWeb umbrella during the recent years. They are featured with a short abstract and their main publications for further details. Part II is devoted to the methods for analysis of complex power systems, where advanced mathematics and measurement techniques are used to tackle the complexity of future power grids. Simulation is the focus of Part III, including fast numerical solvers and flexible co-simulation setups. Managing flexibility and storage is covered in Part IV, both in the electric as well as the thermal energy domain. The topic of vehicle-to-grid received a dedicated part in Part V with a feature on fuel cell vehicles. Part VI, the last one, deals with planning and scheduling under increased uncertainty, stemming from renewable generation.

The PowerWeb consortium will produce this kind of snapshot proceedings in a periodic fashion. For immediate access to its resources, one may consult its webpage http://www.powerweb.tudelft.nl.

Delft, The Netherlands   Prof. Dr. Peter Palensky
June 2018   Chair of PowerWeb, and Head of Intelligent
                                                    Electrical Power Grids, TU Delft

Contents

Part I
Vision and Projects

Chapter 1
PowerWeb Projects

PowerWeb Consortium

Abstract This chapter covers a number of projects that were executed under the PowerWeb umbrella during the recent years.

1.1 Smart Urban Isle

Project Leader: Sabine Jansen

Urban areas are still far from energy neutral. Cities and metropolitan areas consume energy in a number of different ways while producing high levels of CO_2 emissions. Such a large environmental footprint is unsustainable in the long run and solutions are sought that increase local production of renewable energy, improve energy efficiency and reduce CO_2 emissions.

A smart urban isle (SUI) is an area around a bioclimatic building with a smart energy network that creates synergy with other buildings and their energy networks and makes use of the scale advantages for energy solutions. This project specifically examines decentralized renewable energy generation in relation to storage and distribution and electrical mobility, bioclimatological and responsive design of buildings and energy management for controlling energy flows.

The solution is formed by three complementary and integrated energy focused blocks: (1) bioclimatic design system, (2) management platform and (3) mini-networks [1]. A bioclimatic design is achieved by an architectural design with maximum comfort inside the building at minimum energy cost. Since having a bioclimatic public building is not sufficient for energy neutrality, a step forward is made by developing bioclimatic designs in urban areas (bioclimatic urban planning). The management platform collects automatic active measures within the SUI and aims to control and improve the energy efficiency and the energy flows. Finally, SUI mini-networks facilitate the generation, storage and supply of energy within the SUI. Concepts for

PowerWeb Consortium (✉)
Delft University of Technology, Delft, The Netherlands
e-mail: P.Palensky@tudeft.nl
URL: http://www.powerweb.tudelft.nl

© Springer Nature Switzerland AG 2019
P. Palensky et al. (eds.), *Intelligent Integrated Energy Systems*,
https://doi.org/10.1007/978-3-030-00057-8_1

flexible and smart networks integrate the energy demands, storage and renewable energy production, both from buildings as well as from shared energy facilities in the SUI. The aim is to locally balance the energy system as much as possible.

1.2   Optimizing Flexible Energy Use in Industry

Project Leader: Matthijs Spaan

A large portion of renewably generated power is intermittent, uncertain, and uncontrollable. Since a balance between consumption and generation is required at all times, flexibility in electricity consumption is valuable in order to reduce the need for controllable generation on standby (which is typically powered by fossil fuels). A significant body of research has been developed, in addition to some pilot studies, to investigate flexibility of heating, cooling, and electrical vehicle charging in households. However, industry in the Netherlands is using about three times more energy than the residential sector while offering far more promising opportunities to engage in providing flexibility.

The core of the problem to unleash the potential of energy flexibility in industry is to optimize the daily operations at the technical and economic levels. For instance, switching between energy carriers, using and sharing buffers for heat, steam, intermediate products, and varying production can all be considered. In this project the algorithmic techniques to support these decision problems under uncertainty are developed, where the goal is to enable better utilization of existing infrastructure. The stakeholders are operators and business analysts of large industrial plants, policy makers, utility companies and port authorities.

1.3   Heat and Power Systems at Industrial Sites
and Harbors

Project Leaders: Han La Poutré, Peter Palensky

The energy systems are slowly transforming towards the goal of full sustainability. Increasing penetration of renewables in our energy landscape, brings with it challenges such as system reliability in face of highly variable supply. To address such challenges, synergies between energy domains need to be explored and exploited. Among energy domains, heat and electricity sectors have high potential for innovations, especially in times when the idea of electrification of heat sector is gaining traction.

Important locations for the consumption and production of energy are industrial sites (including harbors). In such locations, the different roles of electricity and heat are recognized. Each actor in these environments is associated with higher intensity of generation and consumption than in the domestic environments. Therefore it has a higher impact on the intensity of the dynamics of the energy demand and supply. While in domestic areas peak-shaving techniques is available (e.g. using incentive mechanisms like dynamic pricing and by averaging over many actors), this only holds true to a limited extent for industrial sites and harbors. Here, the actor size and strict schedules of the industrial processes leave little room to adjust.

Despite this, optimizing energy flows to obtain even small efficiency gains still imply significant savings in absolute terms. This makes industrial area an interesting study case to analyse. This project aims at developing solutions for automated power and heat management at industrial sites and harbors with a combination of multiple actors, industrial processes, and external factors. The project is carried out by developing innovative models, simulation systems, agent-based market and coordination mechanisms, and optimization techniques.

1.4 Dynamic Capacity Control and Balancing in the Medium Voltage Grid

Project Leader: Mathijs de Weerdt

It is expected that the future will bring higher grid utilization and will put the electricity grid under increased pressure. Increased deployment of large electric consumer loads such as electric vehicles and electric heating and cooling systems threaten to overload the distribution transformers beyond their rated values. In addition, the volatility of the electricity supply is becoming more prominent due to distributed generation, namely consumer operated solar panels, industrially operated micro-CHP, and medium-voltage wind turbines. Addressing these problems using traditional grid reinforcement and back-up supply generators is considered too expensive and inefficient, and hence, the researchers in this project have looked into techniques to make the electricity network intelligent and future-proof.

This research project aims to develop advanced algorithms for managing the consumption of electricity within the available network capacity. The problem is addressed using planning and control algorithms.

The most notable contribution of this project thus far has been a range of methods to pro-actively control an aggregated fleet of Thermostatically Controlled Loads to overcome temporary grid imbalance or overload [2]. The problem is posed as a planning under uncertainty problem and the solution is sought using multi-agent Markov decision process framework. Since the number of agents is large, the problem is decomposed by decoupling the interactions through arbitrage. From evaluating the adaptive decomposition on both large and small sets, it is found that the newly developed methods are able to return near-optimal solution while remaining scalable. These methods are explained in more detail in Chap. 9.

1.5   Gaming Beyond the Copper Plate

Project Leader: Mathijs de Weerdt

The introduction of heat pumps and electric vehicles may lead to grid congestion, especially during extreme weather conditions when heat pumps run at maximum capacity. In addition, electricity prices will become more weather-dependent as the share of renewable energy in the grid increases. On windy and sunny days when the electricity is cheap, this also increases the danger of significant grid congestion at the distribution level, because flexible loads will all aim to shift to the exact same moment. Current regulations assume the network to be like a copper plate, obliging network operators to facilitate any market transaction at any cost.

In this project, a novel approach to congestion management is proposed, which significantly lowers the social costs when compared to the current solutions such as reinforcing the infrastructure at every point of congestion. The approach relies on a combination of algorithms that can efficiently schedule flexible loads of multiple customers within the capacity constraints of the local energy networks, and new rules for the use of the infrastructure under congestion.

Until now, this project has delivered a couple of major results. First, a method has been developed to use information contained in past wind scenarios to plan loads for upcoming time slots [3, 4]. Second, a new market design has been proposed to allocate network capacity under uncertain conditions [5], and third, a robust and stochastic optimization formulation for wind curtailment has been improved to allow for faster computation and more reliable decisions on which power plants should be running (unit commitment) [6].

1.6   Future-Proof Flexible Charging

Project leader: Mathijs de Weerdt

Electric vehicles (EVs) can significantly contribute to reducing air and noise pollution in metropolitan regions. Charging EVs, however, can put significant strain on local distribution grids. Fortunately, flexibility in the charging process can be exploited to reduce this strain, or to help with balancing. However, optimising trading decisions (in expectation) is a real challenge, because of the uncertainty in electricity prices in the intra-day and balancing markets, whether provided reserve capacity will be used, and the uncertainty in the behavior of the owners of EVs. Furthermore, the complexity of the optimization problem increases significantly if physical limitations of the distribution network also need to be taken into account. And what is the effect of these advanced algorithms on congestion in distribution networks in practice?

In this project smart planning algorithms are developed to coordinate trading flexible power consumption by EVs under such uncertain conditions, considering and comparing stochastic optimization as well as algorithms for sequential decision-making under uncertainty that allow (i) scaling to many loads or generators while (ii) taking network congestion into account. Additionally, the effect of a market for flexibility on the costs for distribution network companies is analyzed through simulation.

We have provided a new stochastic optimization model to make optimal-in-expectation bids for providing reserve power (R2) [7]. The unique contribution is here that the bids include both price and quantity and consider also the prices in the intraday and balancing market. Furthermore, we have shown under which conditions adding a network constraint indeed increases the theoretical complexity of the problem (making it NP-hard) [8]. In current work we are evaluating alternatives to a market for flexibility, for example by introducing advanced tariffs.

1.7 Regional Energy Self-sufficiency

Project Leader: Kees Vuik

For the transition to a more sustainable and autonomous local energy supply system, it is of utmost importance for the local communities to take an active role in the energy supply using local energy sources like wind, solar and biomass. However, due to intermittency and volatility of the wind and solar power, the sustainable local energy supply is exposed to high fluctuations and uncertainty. To ensure high reliability of the energy supply at minimal cost, various aspects such as increased system integration, flexibility, local energy storage and retrieval, and multi-energy systems are investigated and tested.

This project explores how an optimal local energy supply system can be designed with minimal dependence on the national energy grid given the local infrastructure of gas, power, heat and their typical local demand patterns. Alongside the technical aspects of the problem, this research explores the question of adequate regulation for these energy systems and the methods for their efficient operation. Hence, this project aims to provide an optimal technical design along with an institutional design of local energy systems including regulatory framework and governance.

To achieve the desired objective, models and methodologies have to be developed along two directions. First, an optimal overall design of the energy supply system and the corresponding required institutions is to be proposed. The proposed methodology is based on distributed optimization that includes various investment decisions made by different actors. Second, the advances are proposed in simulation and optimization of gas, power and heat flow in their combined networks, for assessing solution uncertainty, sensitivity, and reliability.

1.8 Block Modeling Method

Project Leader: Domenico Lahaye

When a fault happens in an electricity grid, such as short circuit or a collapse of a power grid tower, the voltages and currents in the power network change suddenly, creating so-called transients in power systems. Existing algorithms require large amounts of computational resources, which confines the state-of-the-art industry-used solutions to small scale problems and limited topology considerations.

The aim of this project is to apply recent advances in computational methods for numerical integration of differential equations to the problem of simulating transients in power systems. The goal is to minimize and/or completely remove mathematical operations which require large amount of instructions to complete, achieving high performance as a result.

The main contribution of this project is a method for simulation of non-ideal (lossy) switching in large power system networks [9]. The proposed block modeling method separates static from adjustable parts of the system Jacobian. The main advantage of applying the block modeling method is that the computation of the analytical Jacobian matrix is possible and cheap for any number of arc models and very large power networks. This method has shown significant reduction in computation time when compared with the traditional approach.

1.9 Risk-Based Security Assessment

Project Leader: Domenico Lahaye

Secure grid operation in future electricity networks becomes increasingly complex due to the adoption of new technologies, market deregulation and the gradual increase in consumption. One of the main operational challenges is to improve the security assessment process in order to cope with higher infeed of intermittent generation, lower grid inertia, and novel consumption patterns.

This project studies the risk-based system security concepts in order to develop a dedicated and innovative toolbox to support and improve operational security assessment of transmission system operators.

A major result of this project, reported in [10, 11], is on the inclusion of uncertainty in power system operational security assessment. The proposed model, based on Monte-Carlo framework, captures the stochasticity of the system inputs using the copula theory for the sampling of the system infeeds. Moreover, the model computes the probability of cascading events, while pointing out necessary remedial actions for ensuring security and the incurred associated amount of lost load. The results show that applying simplifications such as assumptions of active-reactive power independence or other approaches that rely on DC power flow approximations could lead to an underestimation of the system risk.

1.10 DC Distribution Smart Grids (DCSMART)

Project Leader: Pavol Bauer

High penetration of distributed energy resources imparts complex behaviors such as stochasticity into the electric power system, which is traditionally being operated with alternating current (AC). Nowadays most devices operate with direct current (DC) internally, and most distributed renewable resources generate power in DC. Moreover, storage components as batteries and supercapacitors have a DC character.

DC distribution grids have the potential to facilitate smart grid applications in a more straightforward way. With DC distribution grids, DC sources and DC loads could be directly connected to a DC bus, eliminating DC/AC and AC/DC conversions. The advantages of such an approach include reduced number of conversion steps which lowers the conversion losses, and elimination of the technologies and markets supporting frequency synchronization. This approach can be facilitated by power electronic converters which provide improved flexibility to the system. In the DC distribution systems the participation of power converter devices is more pronounced than in the AC distribution systems. By deploying appropriate control logic at the power converters, the implementation of ICT- based smart grid concepts, such as active demand side management, can be simplified in the DC distribution grids.

DCSMART is an interdisciplinary project that aims at enabling a straightforward integration of smart grid system technologies, creation of market opportunities and stakeholders adoption through the development and implementation of DC distribution smart grids. The DC distribution smart grids will be based on modular and scalable concepts and will be validated in two demonstration sites, one in the Netherlands and one in Switzerland. Innovative smart markets and smart grid applications such as active demand side management will be implemented.

1.11 Symphony - A Distributed Smart Grid Experiment Platform

Project Leader: Frances Brazier

Integration of renewable energy resources, managing peak demands, increased security and environmental concerns are some of the challenges that electricity networks face in many countries over the world. To answer these challenges, the concept of multi-agent systems that provide means for distributed monitoring, processing and planning has gained momentum. For example, local software agents that act on behalf of a customer in their best interest can be used to rearrange future energy consumption according to pre-defined constraints. Testing of such agent-based solutions prior to on-site deployment is still a challenge since current centralized simulation environments may not capture all the aspects of distributed, large-scale, complex system in which the agents are to operate.

Symphony is a multi-agent distributed experiment and testing platform that supports real-world smart grid experiments with the participation of both simulated and real-world actors [12, 13]. The actors in different physical locations can safely join the experiments while Symphony takes care of interconnection and security concerns for them. The flexible infrastructure provided by Symphony connects participants to distributed services. The participants can then use the platform to experiment with pricing and load balancing for example. Symphony is being developed and deployed in the context of two European Institutes for Innovation and Technology projects across Europe: SES European Virtual Smart Grid Lab and Open SES.

1.12 CIVIS - Cities as Drivers of Social Change

Project Leader: Frances Brazier

By deploying distributed energy resources, citizens are no longer solely energy users but also energy producers. Uncoordinated production of this energy at the community level creates multitude opportunities for improvement and more efficient production, distribution and consumption of energy.

The CIVIS project explores the potential of social networks and communities to significantly reduce energy use and carbon emissions. The role of the project is to enable the opportunities by developing business models for the resulting energy value system and support it with the necessary ICT. More specifically, CIVIS implements a distributed ICT system to (1) manage communities' energy needs, (2) negotiate individual and collective energy service agreements and contracts, (3) raise awareness about the environmental impacts of collective energy use, and (4) allocate energy production resources more efficiently. The project focuses on two pilot neighborhoods located in Trento and Stockholm in close collaboration with energy companies, citizen groups and local administrations.

The project has resulted in several important outcomes thus far. In [14] the role of context in residential energy interventions is reviewed and analyzed. The categorization of all major types of residential energy interventions is performed and used to study their effectiveness in specific contexts: physical (environmental); socioeconomic; cultural; and political and governmental contexts. The analysis resulted in a framework for practitioners and researchers that explicitly includes context when designing successful energy interventions.

This project also made significant contribution in the area of social networking for sustainable energy world. In [15] it was shown that online social networks can be used to form virtual energy communities with shared values such as sustainability and social cohesion. Using an agent-based simulation model it was shown that a large community with occasionally active members form a better predictor for successful energy communities than a smaller community of very active users. This work was extended in [16] by proposing an open source platform for community-oriented user engagement. The community-oriented design is composed of parts that link energy data to energy actions, provide comparisons at different levels, generate dynamic time-of-use signals, offer energy conservation suggestions, and support social sharing.

1.13 ADREM - Adaptive Clustering for Decentralized Resilient Energy Management

Project Leaders: Frances Brazier, Han La Poutré

Distributed energy resources (DER) have seen high incline in deployment of the past couple of decades. They are typically deployed as decentralized, with little or no coordination with other DERs. Hence, addressing the system level objectives in a coordinated fashion using DERs is still an open challenge.

This project focuses on the design of a framework for DER management based on self-optimizing and self-healing clusters of consumers and producers. The participation of the producers and consumers in the clusters is arranged using the negotiated service level agreements (SLA). Clusters are (approximately) autarkic and adaptive. The system is setup in such manner that the cluster membership and SLAs can be (re-)negotiated in response to changes in the environment, the overall energy market, the (forecasted) availability of energy resources, but also participants' forecasts of their own needs and possibilities. This feature results in local, decentralised supply and demand management based on SLAs, reducing complexity at scale while providing the basis for balancing of supply and demand using reconfiguration. In addition, load shedding and system restoration/re-configuration schedules for times of critical need can also be setup using this framework.

The first outcome of this project is a review of the multi-agent-based decentralized energy management issues [17]. This paper also proposes a method for DER management based on dynamic clustering of energy resources for more efficient balancing of supply and demand using SLAs. An extension to this work is reported in [18]. In this paper, static and dynamic virtual clusters are compared. Dynamic reconfiguration is accomplished by changing the time periods of clustering duration. The proposed clustering mechanisms show that decentralized operation of large-scale centralized energy systems is possible if only local information is available.

1.14 Stable and Scalable Decentralized Power Balancing using Adaptive Clustering

Project Leaders: Frances Brazier, Han La Poutré

The energy supply has traditionally followed the variations in the demand. In the future, the demand will have to follow the supply. Having time varying prices is one means to this end. In such approach, markets adapt to the change in operating conditions allowing flexibility on both demand and generation side. Implementing such markets, however, has its challenges. An important challenge is that markets are sometimes highly dynamic, exhibiting even disruptive and chaotic behavior. In addition, the current centralized markets are not designed to adopt to unpredicted events in the grid, such as equipment failures. These factors influence the stability and predictability of the energy system and associated markets. Therefore, there is a need for decentralized and scalable approaches to balance supply and demand of energy.

This project focuses on the design of distributed coordination and market mechanisms for this purpose. Distributed dynamic clusters of synergetic consumers and producers are in the center of the approach. The clusters are designed to coordinate local load balancing for varying periods of time amongst consumers and producers. These time periods are typically substantially longer than the periods considered in the current markets. Local load balancing in clusters thus allows for novel, more reliable solutions for global load balancing and can be used in conjunction with (current or novel) external market mechanisms. In addition, the cluster configuration is dynamic and capable of adapting to changing situations, including network failures.

The most recent results of this project are reported in [19, 20]. On a representative mixed residential and commercial neighborhood in Amsterdam the study of storage coordination is performed. The influence of storage coordination and peak-shaving operation on the neighborhood's energy autonomy and on the peakiness of the power exchanged with the main grid are addressed. Results show that, compared to individual storage operation, coordinated storage operation increases renewable energy utilization by 39% while decreasing the excess energy transferred to the grid by almost threefold and increasing the neighborhood self-sufficiency by 21%. The peak-shaving mode of operation reduces the highest power peak of the year by 55%.

1.15 Solar Powered Bidirectional EV Charger with Smart Charging

Project Leader: Pavol Bauer

Electric vehicles (EV) are considered to be the future mode of transportation. The key drivers for EVs are their higher efficiency and zero tail-pipe emissions. However, EVs are only sustainable if the electricity used to charge them comes from renewable sources and not from fossil fuel-based power plants. The goal of this project is to "Develop a highly efficient, V2G-enabled smart charging system for electric vehicles at workplaces, that is powered by solar energy".

System Design - The system design of the solar EV charging station investigates the best design for the PV system in order to meet the EV charging demands [21, 22]. In spite of the lower solar insolation in the Netherlands, an average of 30 kWh/day is generated by a 10 kWp photovoltaic (PV) system. There is up to five times difference in energy yield between summer and winter. The use of a local storage was found to help in managing the diurnal solar variations but had a negligible effect in overcoming seasonal solar variation.

Power Converter - In this project, a single integrated converter has been built that charges the EV from PV on DC and requires only a single, common inverter for both EV and PV [23]. The charger is bidirectional and can implement vehicle to grid (V2G) where the EV can feed power back to the AC grid. The converter can realize four power flows: PV → EV, EV → Grid, Grid → EV and PV → Grid. Interleaving, silicon carbide (SiC) MOSFETs, SiC Schottky diodes and powdered alloy inductors are used in the converter to achieve both high power density and high efficiency. The EV charger is modularly designed and several 10 kW power modules can be operated in parallel to scale up to higher powers of up to 100 kW easily.

Smart Charging Algorithms - Smart charging refers to the technique of controlling the magnitude and direction of the EV charging power for different applications. In this project, new charging algorithms are proposed that integrate several applications together for charging the EV [24]: PV forecast, EV user preferences, multiplexing of EVs, V2G demand, energy prices, regulation prices and distribution network constraints. For two specific case studies simulated for Netherlands and Texas, the proposed algorithms reduced the net costs in the range of 32–651% when compared to uncontrolled and average rate charging, respectively.

EV-PV Charging Station - The developed EV-PV converter has a much higher peak (95.2% for PV → EV, 95.4% for Grid → EV, 96.4% for PV → Grid) and partial-load efficiency than existing solutions. The power density of the converter is 396 W/l, which is three times that of existing solutions based on Si IGBT technology. The charger is compatible with the CHAdeMO and CCS EV standard. Successfully tests have been carried out with a CHAdeMO compatible Nissan Leaf EV by charging it from PV and feeding power back to the grid via V2G.

1.16 Integration Strategies of Plug-in Electric Vehicles (PEVs): Participation in Ancillary Services

Project Leader: Pavol Bauer

PEVs have a great potential to provide different types of power system ancillary services. The capability of storing energy over long periods of time and the fine-grained instantaneous active power control of the fast-switching converters of PEVs are two attractive features that enable PEVs to be engaged in ancillary services, primary frequency control and voltage support just being some of them.

In the past, the frequency stability analysis of power systems with PEVs has been typically analyzed in the transmission side following a major disturbance. This approach leaves the distribution network neglected and the PEVs invisible to the analysis. This project aims to incorporate the distribution network characteristics including the PEVs behavior and capabilities in the studies of the grid frequency stability.

The major result of this research is reported in [25]. This reference proposes a new model for PEVs based on the participation factor. The proposed model facilitates inclusion of several PEV fleet characteristics such as the minimum desired state of charge, drive train power limitations, constant current and constant voltage charging mode specifications.

1.17 PMU Supported Frequency-Based Corrective Control of Future Power Systems

Project Leader: Marjan Popov

It is anticipated that the increase in renewable generation will drive the power system closer to its stability limits. Particularly, the replacement of conventional power plants with distributed generation will reduce overall inertia of the power system. A power system with small inertia will be more sensitive to disturbances, and if not controlled properly, will be more prone to blackouts.

This project seeks to create a wide-area intelligent system that will improve grid capability to deal with disturbances. The proposed intelligent system provides extensive synchronized information in real-time by using PMU measurements. In addition, it quickly assess the system vulnerability and performs timely corrective control actions. The particular focus of this project is on the design of a new closed-loop corrective control scheme that can be used for elimination of system frequency instability, cascading outages and catastrophic blackouts in existing and future electricity networks.

The main contribution of the project, as of today, has been a computationally efficient and robust algorithm for synchronized measurement technology (SMT) supported online disturbance detection [26]. The novel algorithm is based on the robust median absolute deviation SMT dispersion measure to locate outlier dataset samples. It can be utilized as a pre-step in alternating current (AC) and high voltage direct current (HVDC) protection schemes. It has been shown that a single PMU disturbance-affected measurement is sufficient for accurate detection. In addition, the simulations performed on a personal computer (PC), show that the disturbance detection algorithm is executed in average 0.11 ms per sample dataset window, while the typical stationary wavelet transform disturbance detection technique used for comparison requires 1.18 ms.

1.18 Car as Power Plant

Project Leaders: Zofia Lukszo, Ad van Wijk, Bart de Schutter

Energy and transport systems are more and more interwoven. Both sectors have not only similar environmental goals but they do need each other to achieve these goals. Electric vehicles (EVs), including plugin EVs and fuel-cell electric vehicles (FCEVs), have a huge potential to play an important role in future energy systems. They can be used, when parked, to discharge electricity to the grid. When aggregating the power of a large number of vehicles, they can function as dispatchable power plants. EVs can adapt their charging behaviour to the needs of the power system operator. Similarly, they can act as storage, for example by charging their batteries when there is a surplus of renewable energy.

In 2014 a project "Car as Power Plant" (CaPP) has been started at the Delft University of Technology, where the fuel cell cars are proposed to be used as power plants in a paradigm changing concept, defined by prof. Ad van Wijk [27]. Such fuel cell cars have the potential to create an integrated, efficient, reliable, flexible, clean and smart energy and transport system. The concept is that fuel cell cars do not only contribute to a more efficient and cleaner transportation, but that when parked they can produce electricity more efficiently than the present electricity system and with useful 'waste' products heat and fresh water.

In terms of technology, the energy production system can be envisaged as a fleet of fuel cell vehicles, where cars while parked (over 90% of the time) can produce with the fuel cell electricity, heat and fresh water, which will be feed into the respective grids. From a social perspective the stakeholders directly and indirectly involved in the design, building and operation of such a system, are car park operators, the local power, heat and water distribution companies, gas suppliers, H_2 producers, the equipment, system and software manufacturers but also municipalities, regulators, policy makers and not to forget the car owners/users.

The concept is based on the potential of using FCEVs to replace centralized power plants, and this can be achieved in different ways [28]:

- Using several FCEVs to become part of an energy community system
- Using a parking garage to physically aggregate large numbers of FCEVs
- Using aggregated vehicles to act as back-up power in hospitals.

To facilitate the introduction of such innovative systems as the CaPP combining technical, economic, operational, and social aspects is necessary to obtain a complete understanding of the system. Oldenbroek et al. have shown that realizing a fully renewable energy system based on the CaPP principle is realistic [29, 30]. Alavi et al. have shown with optimal scheduling that it is possible to minimize the electricity import from an external network in a microgrid [31]. Not only insights into the possible consequences of design choices and operational modes provided by advanced modelling and optimisation techniques are important, but at the same time alternative governance structures and institutional considerations, as described in [32], should be taken into consideration, see Fig. 1.1.

To investigate the CaPP system a 100% renewable integrated energy and transport system for a smart city area with 2000 households is described in [33] as an illustrative case study. The operation of this smart city is based on wind, solar, hydrogen, and fuel cell electric vehicles and is inspired by the city of Hamburg in Germany. In this case hydrogen is produced within the urban areas from local surplus solar energy and from shared large scale wind energy, and next hydrogen is transported via tube trailers from

Fig. 1.1 Analysis
framework for the design of
the CaPP system

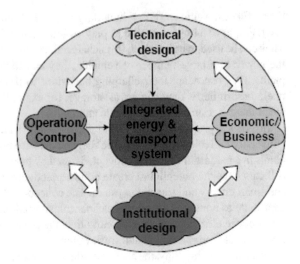

the urban areas to hydrogen fuelling stations, to other hydrogen hubs/consumers, or
to the large scale and shared underground seasonal hydrogen storage.

The whole system has the following major elements, see Fig. 1.2:

- Local Solar Electricity and Hydrogen Production
- Building Electricity Consumption and Smart Electric Grid Control
- Hydrogen Tube Trailer Transport
- Hydrogen Fuelling Stations
- Road Transport with a fleet of road transport FCEVs being passenger cars, vans,
  buses, lorries and trucks
- Large-Scale and Shared Wind Hydrogen Production
- Large-Scale and Shared Seasonal Hydrogen Storage.

Using techno-economic analysis it is shown that such design is technically feasible. To guarantee technical feasibility also the controllability of the system aimed
at maintaining the supply-demand balance as well as minimizing the operational
costs of the FCEVs is taken into consideration. It is also stressed that operation of
such an innovative concept should be accompanied by an institutional analysis and
designing an organizational system structure. To this end, the impact of regulations
and incentives on car drivers' behaviour and CaPP microgrid performance is investigated by means of agent-based modelling. This included an analysis of how the
design of contracts between aggregators and drivers can stimulate drivers to make
their car available for vehicle-to-grid (V2G) operation.

In the transition toward sustainable energy systems, hydrogen may play broader
role than only being used as fuel for FCEVs. Hydrogen is clean and safe energy carrier
that can be used as a fuel in transportation and electricity production or in industry
as feedstock. When renewable electricity is used for hydrogen production we can
talk about solar (or green) hydrogen, which, moreover, produces zero emissions at

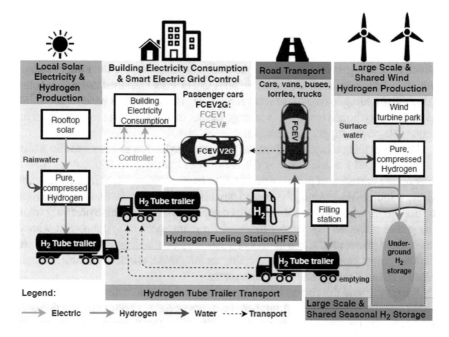

Fig. 1.2 The smart city area of the illustrative case study [33]

point of use. Hydrogen's unique properties make it a potential enabler for the energy transition as it can be used to [34]:

- Enable large-scale, efficient renewable energy integration
- Distribute energy across sectors and regions
- Act as a buffer to increase system resilience
- Decarbonize transport
- Decarbonize industry energy use
- Serve as feedstock using captured carbon
- Help decarbonize building heat and power.

Being convinced that hydrogen will play a key role in the energy transition coordination and incentive policies are needed to encourage deployment of hydrogen solutions and investments.

1.19 CaPP+: Modelling and Design of "Car-as-Power-Plant" Systems in a Real-Life Environment at Shell Technology Centre Amsterdam

Project Leaders: Zofia Lukszo, Ad van Wijk, Samira Farahani

Objective of this project is to model and design real-life systems for the 'Car as Power Plant' (CaPP) concept. The challenging CaPP-concept, initiated by prof. Ad van Wijk, is researched at TU Delft through the URSES project Car as Power Plant to investigate utilization of automotive fuel cell systems as stationary power production units during non-driving hours (typically at home or in a car park during office hours). When the car is parked, the fuel cell unit delivers power back to the grid, to balance the electricity grid, decrease peak demands or serve as base load power generator. It converts hydrogen (produced from renewable sources such as wind or solar) into electricity. Hydrogen is used as a storage means to decouple (in time and distance) renewable power supply and demand.

Using the fuel cell system of the car as power production unit has specific advantages. Fuel Cell Electric Vehicles (FCEVs), with hydrogen as fuel, can be used to support the operation of power systems with a large participation of RES (Renewable Energy Sources) be offering the needed flexibility. Moreover, the fuel cell system can be used more efficiently as a decentralized energy system (available precisely where the power consumer is) without the need for extra space or systems, and without centralized power production units. CaPP has the potential to replace electricity production power plants worldwide, creating an integrated, efficient, reliable, flexible, clean and smart energy and transport system.

We now extend this research with modelling and designing the system in real-life environments, where a limited number of cars are operated as power production units when the cars are parked. To this end, we will utilize a well-controlled environment at Shell Technology Centre Amsterdam (STCA). Here, we focus on how to integrate the CaPP-system into life buildings and the energy micro-grids of the STCA location. For this purpose, we will study three scenarios: (1) all electric; (2) hydro-electric; (3) combining scenario one and two. In the first two scenarios, we study extreme cases of having connection to only the electricity grid (scenario I) and only the hydrogen grid (scenario II); in the last scenario, we study the best combination of these two scenarios, both from the energy efficiency and from the cost efficiency points of view.

1.20   Aquifer Thermal Energy Storage Smart Grids (ATES-SG)

Project Leaders: Tamás Keviczky, Jan Kwakkel, Theo Olsthoorn

Approximately 2000 Aquifer Thermal Energy Systems (ATES) are installed all over the Netherlands. It is expected that this number will increase to 20.000 within 10 years leading to a reduction of 11% in CO_2 emissions, along with estimated savings of 4 billion euros in the coming 30 years. The energy saving potential of ATES at a global scale is even bigger. However, the current performance of ATES is still under expectations and the projected efficiency remains yet to be reached.

The disappointing current contribution of ATES to energy efficiency is mainly due to the present operation and regulation practices that cannot cope with uncertainties in aquifer characteristics, insufficient interaction of neighboring systems, and variability in weather conditions. ATES interact via the groundwater aquifer in a way comparable to how distributed sources and sinks of electricity are interacting via the electricity grid. In ATES, however, the links are time-varying by nature and plagued by uncertainty in connection due to the absence of models and lack of cooperation with the nearby systems.

Distributed Model-based Predictive Controllers (D-MPC) promise significant benefits in operation of ATES by ensuring near-optimal control policies in ATES grids while enforcing critical operating constraints. However, stochastic uncertainties with probabilistic time-varying constraints have never been incorporated in the design of such a distributed control network. This research sets out to deliver a proof-of-concept for the potential of D-MPC in the development of ATES systems into ATES Smart Grids under uncertainty, as a part of the NWO-sponsored URSES program.

The first results of this research are reported in [35, 36]. In these references, we develop a novel large-scale stochastic hybrid dynamical model to predict the dynamics of thermal energy imbalance in smart thermal grids consisting of building climate comfort systems with hourly-based operation and ATES as a seasonal energy storage system. We formulate a finite-horizon mixed-integer quadratic optimization problem with multiple chance constraints at each sampling time, and develop a computationally tractable framework to approximate its solution. The results show efficient use of ATES and point towards a general trade-off between individual and collective ATES performance.

In our most recent work, we provide a technique to decompose the large-scale scenario program underlying the decision-making problem into distributed scenario programs that exchange a certain number of scenarios with each other in order to make local decisions. We show that such a decomposition technique can be applied to large-scale linear systems with both private (local)and common uncertainty sources. This yields a flexible and practical plug-and-play distributed scenario MPC framework. These results are currently being applied in Amsterdam as part of a pilot implementation project supported by the NWO URSES+ program.

1.21 Jouw Energie Moment

Project Leader: Charlotte Kobus

Demand response has been widely accepted as the means for obtaining greater energy efficiency and adopting higher levels of renewable generation. However, a significant practical challenge lies in engaging energy consumers to participate in providing flexibility to the grid. The main aspect of this challenge is to bring the users to relate to the energy needs of our society and to empower them to become proactive in ensuring sustainable energy future.

In this project, we aim to understand whether and when residential consumers are able and willing to change their electricity demand to match scarce supply conditions. Specifically, we are interested in what design (e.g. interface design, smart appliances) can do to contribute to habit formation so that the behavioral changes will last over time.

Thus far, the project has completed a qualitative study on a small set of households who were asked to 'wash when the sun is shining' [37, 38]. The researchers are currently analyzing the results of this field study amongst 250 households in Breda and Zwolle, who received solar panels, an EMS, a dynamic pricing tariff, and a smart washing machine to shift the electricity demand of their washing machine over time. The point of interest is to understand if these households are changing electricity demand to match supply conditions and why. In addition, the interaction of the participants with the provided technology is observed so that new design guidelines can be proposed to enhance this interaction.

1.22   Platform Wars for Socially Responsible Smart Grids

Project Leader: Geerten van de Kaa

Complex systems such as the smart grid are technically possible to realize but they are not implemented on a large scale. One of the underlying reasons is that generally accepted compatible interfaces (or standards) are lacking with which the components of such systems can be interconnected. The problem is not that there are no interfaces. In fact, there are many interfaces which are competing in 'standards battles'. My research focuses amongst others on factors that affect the outcome of standards battles.

I have developed a framework consisting of 29 factors for standard dominance which I have applied to various cases of standards battles for complex systems. Factors include firm's resources and strategies as well as a standard's technical characteristics such as the compatibility that it enables. Also, market mechanisms such as network effects are taken into account. I have tested the completeness and relevance of the framework and I have explored the extent to which weights can be assigned to factors. It appears that the inter-organizational network of stakeholders that are supporting the standard is essential for achieving success. In some papers I have explored several elements of the inter-organizational network of stakeholders on standard success.

To fully understand standards battles, a broader approach is needed than just focusing on a firm in its economic environment. Functional and ethical consumer values also need to be considered and may lead to increased standard acceptability and thus to standard selection. This novel notion is and was explored in two NWO-MVI projects and the Horizon2020 project IAMRRI.

1.23 Integrated Community Energy Systems (ICES)

Project Leader: Paulien Herder

The local energy landscape is radically transforming in front of our eyes. More and more individuals and organizations that have historically been identified as consumers of energy are becoming prosumers by generating their own electricity locally thanks to the adoption of suitable policies at the national level, cost reduction of renewables, and technology developments in ICT and energy domains. More energy balancing options are feasible when the prosumers cooperate. However, increasing distributed generation due to advent of renewable energy technologies as well as increasing electricity demand due to electrification of new sectors (e.g. electric vehicles, heat pumps) provide additional challenges to maintain integrity of the electricity grid.

As a part of the solution, innovative ways for local matching of demand and supply in the form of ICES are emerging in the energy systems. They refer to multiple approaches for supplying local communities with required energy from high-efficiency co-generation or tri-generation as well as from renewable energy technologies coupled with innovative energy storage solutions including electric vehicles and energy efficiency demand-side measures. Significant benefits associated with integrating these technologies among others flexibility and robustness could assist further advancements in the smart grids.

This research aims at investigating the value of ICES and at providing the recommendations for institutional design to clearly define the role and responsibilities of different actors involved as well as to institutionally embed these systems in the existing energy systems. In [39] issues and trends shaping the ICES are reviewed. In [40] a model-based framework is presented that assesses the value of ICES for the local communities. A distributed energy resources-consumer adoption model (DER-CAM) is used to assess the value of an ICES in the Netherlands.

1.24 Reliability Evaluation of Power Systems with Integrated Large-Scale Offshore Wind Energy

Project Leaders: José Rueda Torres, Mart van der Meijden

In order to transport offshore wind energy to the main load centers in the Netherlands and the other European countries, the Extra-High Voltage (EHV) transmission network needs to be reinforced. Since the society prefers to apply underground cables rather than overhead lines, EHV cable systems are currently installed in the Dutch transmission network in the Randstad380 project. These new grid developments like Extra-High Voltage (EHV) underground cable systems and networks for offshore wind energy can significantly influence the reliability of the power system. It is therefore of interest to study the impact of these developments on the reliability of the grid in order to secure the electricity supply in the future.

In this research, the reliability of the Dutch EHV transmission network is studied using an approach that can be applied to other systems as well. Probabilistic approaches are compared with deterministic criteria in order to find the best method to assess the reliability of transmission networks.

The results of the research are reported in several references. The tradeoff between network redundancy and generation reserve for combined onshore-offshore transmission networks is analyzed in [41]. In [42], an analysis of reliability of transmission links consisting of overhead lines and underground cables is investigated. The tradeoff between net present value and reliability for offshore transmission links is investigated in [43], while reliability of EHV and underground cables in the case of the Netherlands is considered in [44]. The results of this research are published in a Ph.D. dissertation [45].

1.25   PowerParking: Integration of Solar Photovoltaics with Aggregated Electric Vehicles at Large Parking Facilities

Project Leader: Ad van Wijk

For locations where large numbers of electric vehicles (EVs) are parked, many of which need to be simultaneously charged, there can be both extremely high electricity consumption as well as high peak demand. However, the large battery capacity provided by the aggregation of these parked vehicles, if managed optimally, can also be used for reducing the imbalance caused due to the intermittency and variability of solar photovoltaics within the same microgrid.

Ideally, the energy management system designed for such a microgrid would reduce the overall and peak energy demand from the vehicles, enable the increased use of lower cost local photovoltaic (PV) energy and still deliver electricity to the vehicles at an adequately high state of charge when they are needed by the user.

The PowerParking project aims to investigate the technical feasibility and potential business cases for such an integrated solar powered electric vehicle charging system for large parking facilities. Systems for both long term parking, such as those in airports, and workplace parking, such as those in employment centres, will be analysed for suitable infrastructure and energy management. Given that EVs and solar PV are both technologies that work with direct current (DC), this project aims at enabling their integration through a DC microgrid. This will be done in order to avoid unnecessary inversion-rectification electricity conversions and increase the overall efficiency of energy flow within the system.

The PowerParking project involves cooperation between multiple industrial partners together with local government, real estate developers and academia.

A prototype of the modular and scalable carport infrastructure will be tested and validated at the Green Village testing grounds at the Delft University of Technology, after which larger scale trials will be undertaken at the new airport and businesspark in Lelystad, the Netherlands. The project is in line with efforts to increase the use of locally produced sustainable energy and low emission mobility in new infrastructure in the Dutch province of Flevoland.

1.26   CESEPS: Co-evolution of Smart Energy Products and Services

Project Leaders: Ad van Wijk, Carla Robledo

The aim of the project is to support the development of smart energy products and services for local smart grids that better respond to the demands and concerns of all stakeholders in terms of performance, cost, reliability, safety and robustness, sustainability and energy-efficiency, and end-users' comfort [46]. The research in this project is focused on comparative validation of technologies and concepts of existing demonstrations and the further development of new innovative energy products and services for the present and medium-term using a co-evolutionary approach. The project involves research teams from the Netherlands and Austria, The Dutch team is composed of four universities, namely University of Twente, TU Delft, Wageningen University and Utrecht University and an international company based in the Netherlands, DNV GL. The Austrian team comprises the Technical University of Graz, the Austrian Institute of Technology, and the European Sustainable Energy Innovation Alliance.

To realize these objectives and to achieve these goals, the project will perform a comparative validation of smart grid technologies and concepts in more than four existing demonstration projects in the Netherlands under the umbrella of the Smart Energy Collective of DNV GL, such as PowerMatching City, Your Energy Moment, Heerhugowaard, Lochem, Gorinchem and a new pilot at Ameland and in six ongoing pilots Austria called E-mobility on Demand, PlanGridEV, iWPP-Flex, EcoGrid EU, hybrid-VPP4DSO, IGREENGrid, and others. Adding to existing smart grid pilots, both Dutch and Austrian consortium members will develop innovative technological concepts for e-vehicles with fuel cells [47], solar charging and other charging solutions within the framework of the Green Village (TUD), Living Lab (UT), Energy storage (TU Graz), COTEVOS (AIT), DCgrids (TUD), Vehicle2Grid Utrecht (UU) through simulation and experimental laboratory set ups.

1.27   Network Science Applications to Power Grids

Project Leaders: Piet Van Mieghem, Fernando A. Kuipers

In current practice, power system analysts carry out the security assessments of power grids mainly via flow-based simulations. Under certain demand and generation profiles, analysts use the nonlinear AC and/or linearised DC power flow analyses to estimate the steady-state operation of the power grid. Although N-1 and N-2 contingency analyses may be possible from the computational point of view, evaluating scenarios where more than three components fail at the same time requires substantial computational time due to the complexity of the simulation models. However, various outages do occur and could result in very large blackouts. Thus, additional complementary measures to traditional flow-based assessments are needed to analyse and understand the subtle behaviour of power grids.

The power grid is now one of the most complex technological networks. The interactions between a large number of components govern the global flow behaviour and the spread of failures. This complex nature of power grids and its underlying structure make it possible to analyse power grids relying on network science. The applications of network science on power grids have shown the promising potential to capture the interdependencies between components and to understand the collective emergent behaviour of complex power grids [48].

This project focuses on the increasing need of reliable power grids and the merits of network science on the investigation of power grids. In this context, relying on network science, we model and analyse the power grid and its near-future challenges in terms of line removals/additions, malicious attacks, cascading failures, and renewable integration. The project provides tools to investigate link failure/addition, critical asset identification and targeted attacks, cascading failures, and wind power integration [49–52]. The developed concepts in this project extend the state of the art in the applications of network science on power grids and (i) can be the interest of researchers in the field, (ii) can support grid operators in analysing the vulnerability of their network to the current and the near-future challenges, and (iii) can assist decision makers and investors with the planning for the future trends in power grids.

1.28   Enabling Peer-to-Peer Energy Trading by Leveraging Prosumer Analytics (P2P-TALES)

Project Leader: Sergio Grammatico

Small energy actors - especially small businesses and residential end-users - with distributed energy generation and storage units connected behind the meter (prosumers) are currently at the centre of the energy transition. Much research has focused on their engagement, either in terms of directly controlling their consumption or by exposing them to time-varying price signals reflecting the power system or wholesale electricity market conditions. However, centralized ways of stimulating the prosumers' activation have not been proven successful due to the compromise in the end-users' comfort and the limited economic incentives provided.

With the aim to increase the engagement of small prosumers, P2P-TALES proposes the use of data-driven knowledge extraction techniques in order to identify clusters of prosumers with similar or dissimilar energy-related behavioural characteristics, preferences and flexibility potential and create incentive schemes that reflect their financial and non-monetary interests.

P2P-TALES will strive to (1) cluster the behaviour of energy prosumers by applying machine learning methods to users' big data sets, (2) conceive distributed trading protocols among prosumers by developing distributed learning methods for multi-agent equilibrium problems, and (3) assess the impact of trading mechanisms on the distribution network and in turn define system services that can be offered by prosumers so that integration of the cyber and physical systems can be achieved.

1.29 Distributing Load Flow Computations Across TSO and DSO Boundaries

Project Leaders: Domenico Lahaye, Miloš Cvetković

Operating future power systems requires TSOs and DSOs to intensity collaboration. ENTSO-e facilitates this transition at the TSO level by defining standards for sharing of information. This project promotes a framework for the joint AC load-flow computation of a power system shared among various owners. The distinct feature the framework proposed is the small amount of information that needs to be shared to obtain a convergence.

We solve the AC load flow equations using a Newton–Krylov method. This implies that the Jacobian linear systems at each Newton step are solved by a sequence of Krylov subspace iterations. To distribute computations over the set of owners, we adopt a block preconditioner in which the blocks group the nodes of same ownership. For the implementation we take advantage of the Portable Extensible Toolkit for Scientific Computations (PETSc) and the Message Passing Interface (MPI) for the parallel communication.

Our results show that distributed load flow computations can indeed be performed with owners sharing limited information only. The same power flow solution as in a monolithic, share-all information approach is obtained at the same number of Newton iterations. The convergence of the Krylov subspace iterations can significantly be accelerated by allowing a small number of overlapping nodes in the network decomposition. This framework can be extended to other type of network computations for which the load flow kernel serves as a building block. This work is therefore expected to significantly stimulate the interaction between TSOs and DSOs.

1.30   Capturing the Societal Value of Smart Energy Systems

Project Leaders: Paulien Herder, Rolf Künneke, Geerten van de Kaa, Emile Chappin

Smart energy systems comprise the ICT enabled low voltage electricity grids, the smart meters as 'gatekeepers' and the advanced technologies inside the homes that are able to monitor and control the use of energy. Their primary purpose is to increase the electricity grid's ability to cope with unanticipated fluctuations in electricity generation at both side of the tradition electricity supply chain from e.g. wind and solar energy. However, there are serious ethical concerns related to their use, concerning privacy, security, reliability, or affordability, which may ultimately jeopardize their deployment. Traditional social cost benefit analysis is not equipped to properly address these moral values, because it restricts analyses to monetary terms and fails to argue why certain values need to be taken into account and how they can be traded off against each other. To secure the deployment of, among others, renewable energies, there is a strong urgency to take moral values into consideration.

The aim of the project is to identify different technical and institutional designs for smart energy systems that are morally more acceptable. We use the capability approach that was proposed by Sen and Nussbaum, which is well suited to identify embedded values related to energy systems. This approach allows to go beyond traditional financial and utilitarian methods for evaluations of large infrastructures [53]. We rely on the field of institutional economics and hence describe smart energy systems in both technical and institutional terms. This allows to explore the dependency between system acceptability and related regulations. Finally, we describe smart energy systems as complex socio-technical system in which technological innovations co-evolve with social developments. Agent-based modelling is used to explore the dynamic relationship between system acceptability and the set of (emerging) values at stake for this system. Empirical case study research is used to study the impact of different socio-technical design on system acceptability and acceptance.

Project results include a comprehensive literature review and an empirical analysis of public debates on smart grids to uncover relevant smart grid values and their interrelations [54]. In [55], an approach using probabilistic topic models is proposed to review the literature across multiple disciplines addressing latent topics, such as moral values. The project is funded by the Dutch Organization for Scientific Research (NWO) and runs from December 2015 until January 2020.

1.31 TSO2020 - Electric Transmission and Storage Options for 2020

Project Leaders: José Rueda Torres, Mart van der Meijden

The transition towards a renewable energy supply takes place in various sectors. To facilitate this transition, a synergy is created between electricity and hydrogen gas. In TSO2020, the possibilities to install an electrolyser in the northern part of the Netherlands are studied. Hydrogen production can then be combined with a large inflow of electricity from offshore wind farms and submarine interconnections like COBRAcable. In this way, the excess of electrical energy can be converted into hydrogen, to be used by industries and transport, or to be converted to syngas and injected into the natural gas network.

TSO2020 consists of a cost-benefit analysis, a pilot in the Netherlands and an analysis of the scale-up possibilities for mass production. Delft University of Technology is appointed to study the electrical aspects, thereby concentrating on the impact of electrolysers on the stability of the power system. The possibilities for ancillary services (i.e. frequency control, voltage control and congestion management) and the development of control strategies are considered as well. The analysis will be performed using the Real Time Digital Simulator (RTDS). Currently, the model of the electrolyser has been developed [56]. Furthermore, an initial investigation of the possibilities to provide ancillary services has been performed [57, 58]. The possibilities for ancillary services are promising, as electrolysers are able to change their power consumption relatively fast, thereby contributing to the frequency and voltage stability of the power system. The work at Delft University continues with the analysis of the behaviour of a (small and large) electrolyser in a larger transmission network, further investigation of the possibilities for ancillary services and the development of control strategies. More information can be found on the TSO2020 project website [59].

1.32 ERIGrid - European Research Infrastructure Supporting Smart Grid Systems Technology Development, Validation and Roll Out

Project Leaders: Peter Palensky, Arjen van der Meer

ERIGrid is a Horizon 2020 project funded and supported by the European Commission. ERIGrid aims at integrating and enhancing the necessary research services for analysing, validating and testing Smart Grid configurations. It takes a holistic, cyber-physical systems based approach by integrating 18 European research centres and institutions to jointly develop common methods, concepts and procedures. The overall objective of ERIGrid is to optimize the use and development of

Smart Grid research infrastructures in Europe. The benefit to European scientific community is that it remains at the forefront of the research advancement while helping industry to strengthen its knowledge base and validate its technological know-how. ERIGrid aims to achieve its objectives by fostering innovation in three areas:

Networking Activities - Validating Smart Grid technologies and developments is a task that requires a holistic view on the overall process. The entire domain spectrum of future Smart Grids has to be taken into consideration. Besides the technical components, such as the grid infrastructure, storage, generation, consumption, etc., Smart Grid also comprises customers, markets, ICT, regulation, governance, and metrology. A holistic approach demands to integrate all prospective R&D sites and stakeholders, e.g. hardware/software simulation labs as well as academic and industrial research. With this networking activity the next generation of researchers and engineers are being trained.

Joint Research Activities - A core activity in ERIGrid is the provision of distributed and integrated research infrastructure which is capable to support the validation and testing of Smart Grid configurations. Overall, 21 installations provided by the consortium members are available for trans-national access projects.

Trans-National Activities - With the trans-national access (TA) to the integrated research infrastructure of the ERIGrid members, European industrial and academic researchers in the Smart Grid domain can design, develop and test new technologies and concepts. This access is funded by the project and is therefore offered free of charge. For each TA, two calls for proposals per year are organized by the consortium. The consortium publicizes a description of each facility on the project website in order to provide technical information and the central entry point to potential users. The projects submitted by industrial and academic researchers have to tackle challenging scientific and complex technical impacts.

Some Networking Activities and Joint Research Activities have already finished with productive developments and interesting results especially in the field of co-simulation in Cyber Physical Energy Systems (CPES). In [60], an effective way has been described for integrating various research infrastructures in ERIGrid for testing and validation of CPES, while [61] proposes a new approach of synching Power Hardware in Loop with co-simulation for holistic validation of CPES. In [62] a novel method for modelling and defining test specifications has been introduced. Further work will study upscaling of existing CPES systems and analyse the complexities arising due to it.

1.33 EASY-RES - Enable Ancillary Services by Renewable Energy Sources

Project Leaders: Miloš Cvetković, Peter Palensky

The stability and security of the traditional electrical power system are largely based on the inherent properties of synchronous generators (SGs). Such properties are the grid-forming capability, the inertia, the damping of transients, and the provision of large currents during faults. The growing penetration of converter-interfaced and inertia-less Distributed Renewable Energy Sources (DRES) will eventually replace dispatchable SGs and increase power volatility, causing large frequency deviations and voltage regulation problems. The previously proposed solutions to resolve this problem are an increase of SG spinning reserves, the grid reinforcement and the use of central electric energy storage systems, but these centralized solutions come with the high cost.

EASY-RES offers unified bottom-up approach and develops novel control algorithms for all converter interfaced DRES, to enable them to operate similarly to conventional SGs, providing to the grid inertia, damping of transients, reactive power, fault ride through and fault-clearing capabilities, and adaptable response to primary and secondary frequency control. These new functionalities will be transparent to all grid voltage levels.

The key objectives of this project are to increase the robustness of the power system towards abrupt frequency changes by introducing virtual inertia and damping in DRES. Contribute to the stability of the grid by providing frequency-dependent active power. Increase the renewable energy penetration levels at both low voltage (LV) and medium voltage (MV) levels, while avoiding investments for grid reinforcement. The project aims to make the DRES more grid-friendly by (i) reducing the short-term electric power fluctuations at both DRES and HV/MV substation level, and by (ii) introducing active harmonics filtering to each DRES converter. The long-term grid security will be preserved even under very large DRES penetration, by reducing reserve requirements after fault recovery. Finally, the project aims to develop viable business models for all the stakeholders by (i) proposing new metrics for the quantification of the various ancillary services, and by (ii) evaluating the economic cost and benefit of all developed ancillary services.

References

1. S. Jansen, R. Bokel, A. van den Dobbelsteen, Smart urban isle. Rumoer, Period. Voor De Bouwtechnologie **62**, 8–15 (2016)
2. F. de Nijs, M.T.J. Spaan, M. de Weerdt, Best-response planning of thermostatically controlled loads under power constraints, in *Proceedings of the Twenty-Ninth AAAI Conference on Artificial Intelligence* (2015), pp. 615–621

3. E. Walraven, M.T.J. Spaan, Planning under uncertainty for aggregated electric vehicle charging with renewable energy supply, in *Proceedings of the 22nd European Conference on Artificial Intelligence* (2016), pp. 904–912

4. E. Walraven, M.T.J. Spaan, Accelerated vector pruning for optimal POMDP solvers, in *Proceedings of the 31st AAAI Conference on Artificial Intelligence* (2017), pp. 3672–3678

5. R. Philipsen, M. de Weerdt, L. de Vries, Auctions for congestion management in distribution grids, in *13th International Conference on the European Energy Market* (2016), https://pure.tudelft.nl/portal/files/4604595/eem_dgcm_PDFEXPRESS.pdf

6. G. Morales-Espana, A. Lorca, M.M. de Weerdt, Robust unit commitment with dispatchable wind power. Electr. Power Syst. Res. **155**, 58–66 (2018)

7. K. van der Linden, M. de Weerdt, G. Morales-Espana, Optimal non-zero price bids for EVs in energy and reserves markets using stochastic optimization, in *Proceedings of the 15th International Conference on the European Energy Market (EEM)* (2018)

8. M. de Weerdt, M. Albert, V. Conitzer, K. van der Linden, Complexity of scheduling charging in the smart grid, in *Proceedings of the Twenty-Seventh International Joint Conference on Artificial Intelligence* (2018)

9. R. Thomas, D. Lahaye, C. Vuik, L. van der Sluis, Simulation of arc models with the block modelling method, in *International Conference on Power System Transients* (Cavtat, Croatia, 2015)

10. M. de Jong, G. Papaefthymiou, D. Lahaye, K. Vuik, L. van der Sluis, Impact of correlated infeeds on risk-based power system security assessment, in *2014 Power Systems Computation Conference* (2014), pp. 1–7

11. M. de Jong, G. Papaefthymiou, P. Palensky, A framework for incorporation of infeed uncertainty in power system risk-based security assessment. IEEE Trans. Power Syst. **99**, 1–1 (2017)

12. Z. Genç, M. Oey, F. Brazier, Monitoring stakeholder behaviour for adaptive model generation and simulation: a case study in residential load forecasting, in *2015 IEEE/WIC/ACM International Conference on Web Intelligence and Intelligent Agent Technology (WI-IAT)*, vol. 2 (2015), pp. 29–34

13. Z. Genç, M. Oey, H. van Antwerpen, F. Brazier, *Dynamic Data-Driven Experiments in the Smart Grid Domain with a Multi-agent Platform* (Springer International Publishing, Cham, 2016), pp. 121–131

14. S. Šćepanović, M. Warnier, J.K. Nurminen, The role of context in residential energy interventions: a meta review. Renew. Sustain. Energy Rev. **77**, 1146–1168 (2017)

15. Y. Huang, M. Warnier, F.M.T. Brazier, D. Miorandi, Social networking for smart grid users - a preliminary modeling and simulation study, in *Proceedings of the 12th IEEE International Conference on Networking, Sensing and Control (ICNSC'15), IEEE* (Taipei, Taiwan, 2015), pp. 438–443

16. Y. Huang, H. Hasselqvist, G. Poderi, S. Šćepanović, F. Kis, C. Bogdan, M. Warnier, F. Brazier, Youpower: an open source platform for community-oriented smart grid user engagement, in *Proceedings of the 14th IEEE International Conference on Networking, Sensing and Control* (2017)

17. F. Brazier, H. La Poutre, A.R. Abhyankar, K. Saxena, S.N. Singh, K.K. Tomar, A review of multi agent based decentralised energy management issues, in *2015 International Conference on Energy Economics and Environment (ICEEE)* (2015), pp 1–5

18. S. Čaušević, M. Warnier, F.M.T. Brazier, Dynamic, self-organized clusters as a means to supply and demand matching in large-scale energy systems, in *Proceedings of the 14th IEEE International Conference on Networking, Sensing and Control* (IEEE, New York, 2017), pp. 568–573

19. M. Warnier, N. Voulis, F. Brazier, The case for coordinated energy storage in future distribution grids, In *Proceedings of the International Electricity Conference and Exhibition (CIRED)* (2017)

20. M. Warnier, N. Voulis, F. Brazier, Storage coordination and peak-shaving operation in urban areas with high renewable penetration, in *Proceedings of the 14th IEEE International Conference on Networking, Sensing and Control* (2017)

21. G.R. Chandra Mouli, P. Bauer, M. Zeman, Comparison of system architecture and converter topology for a solar powered electric vehicle charging station, in *2015 9th International Conference on Power Electronics and ECCE Asia (ICPE-ECCE Asia)* (2015), pp. 1908–1915
22. G.R. Chandra Mouli, P. Bauer, M. Zeman, System design for a solar powered electric vehicle charging station for workplaces. Appl. Energy **168**, 434–443 (2016)
23. G.R. Chandra Mouli, J.H. Schijffelen, M. van den Heuvel, M. Kardolus, P. Bauer, A 10 kW solar-powered bidirectional EV charger compatible with chademo and COMBO. IEEE Trans. Power Electron. (2018)
24. G.R. Chandra Mouli, M. Kefayati, R. Baldick, P. Bauer, Integrated PV charging of EV fleet based on dynamic prices, V2G and offer of reserves. IEEE Trans. Smart Grids (2017)
25. S. Izadkhast, P. Garcia-Gonzalez, P. Frías, An aggregate model of plug-in electric vehicles for primary frequency control. IEEE Trans. Power Syst. **30**(3), 1475–1482 (2015)
26. M. Naglič, L. Liu, I. Tyuryukanov, M. Popov, M.A.M.M. van der Meijden, V. Terzija, Synchronized measurement technology supported AC and HVDC online disturbance detection. Electr. Power Syst. Res. **160**, 308–327 (2017)
27. A. van Wijk, L. Verhoef, *Our Car as Power Plant* (IOS Press BV, Amsterdam, Netherlands, 2014)
28. Z. Lukszo, E. Park Lee, Demand side and dispatchable power plants with electric mobility, in: *Smart Grids from a Global Perspective*, ed. by A. Beaulieu, J. de Wilde, J. Scherpen. (Springer, Berlin, 2016)
29. V. Oldenbroek, L.A. Verhoef, A.J.M. van Wijk, Fuel cell electric vehicle as a power plant: fully renewable integrated transport and energy system design and analysis for smart city areas. Int. J. Hydrog. Energy **42**, 8166–8196 (2017)
30. V. Oldenbroek, S. Alva, B. Pyman, L.B. Buning, P.A. Veenhuizen, A.J.M. van Wijk, Hyundai ix35 fuel cell electric vehicles: degradation analysis for driving and vehicle-to-grid usage, in *The 30th Electric Vehicle Symposium* (Stuttgart, Germany, 2017)
31. F. Alavi, E.P. Lee, N. van de Wouw, B. De Schutter, Z. Lukszo, Fuel cell cars in a microgrid for synergies between hydrogen and electricity networks. Appl. Energy **192**, 296–304 (2017)
32. E.P. Lee, Z. Lukszo, P. Herder, Aggregated fuel cell vehicles in electricity markets with high wind penetration, in *Proceedings of IEEE International Conference of Networking, Sensing and Control, (ICNSC)* (Zhuhai, China, 2018)
33. S. Farahani, R. van der Veen, V. Oldenbroek, F. Alavi, E.H.P. Lee, N. van de Wouw, A. van Wijk, B. De Schutter, Z. Lukszo, Hydrogen-based integrated energy and transport system. Submitted to IEEE SMC Trans. (2018)
34. Hydrogen Council. How hydrogen empowers the energy transition (2017), http://hydrogencouncil.com/
35. V. Rostampour, M. Jaxa-Rozen, M. Bloemendal, T. Keviczky, Building climate energy management in smart thermal grids via aquifer thermal energy storage systems. Energy Procedia **97**, 59–66 (2016). European Geosciences Union General Assembly 2016, EGU Division Energy, Resources and the Environment (ERE)
36. V. Rostampour, T. Keviczky, Probabilistic energy management for building climate comfort in STGs with seasonal storage systems. Appear Trans. Smart Grid (2018)
37. B.A.C. Kobus, R. Mugge, J.P.L. Schoormans, Washing when the sun is shining! how users interact with a household energy management system. Ergonomics **56**(3), 451–462 (2013). PMID: 23009607
38. B.A.C. Kobus, E.A.M. Klaassen, R. Mugge, J.P.L. Schoormans, A real-life assessment on the effect of smart appliances for shifting household electricity demand. Appl. Energy **147**, 335–343 (2015)
39. B.P. Koirala, E. Koliou, J. Friege, R.A. Hakvoort, P.M. Herder, Energetic communities for community energy: a review of key issues and trends shaping integrated community energy systems. Renew. Sustain. Energy Rev. **56**, 722–744 (2016)
40. B.P. Koirala, J.P. Chaves Ávila, T. Gómez, R.A. Hakvoort, P.M. Herder, Local alternative for energy supply: performance assessment of integrated community energy systems. Energies **9**(12), 981 (2016)

41. B.W. Tuinema, J.L. Rueda, M.A.M.M. van der Meijden, Network redundancy versus generation reserve in combined onshore-offshore transmission networks, in *2015 IEEE Eindhoven PowerTech* (2015), pp. 1–6

42. B.W. Tuinema, J.L. Rueda, L. van der Sluis, M.A.M.M. van der Meijden, Reliability of transmission links consisting of overhead lines and underground cables. IEEE Trans. Power Deliv. **31**(3), 1251–1260 (2016)

43. R.E. Getreuer, B.W. Tuinema, J.L. Rueda, M.A.M.M. van der Meijden, Multi-parameter approach for the selection of preferred offshore power grids for wind energy, in *2016 IEEE International Energy Conference (ENERGYCON)* (2016), pp. 1–6

44. N. Kandalepa, B.W. Tuinema, J.L. Rueda, M.A.M.M. van der Meijden, Reliability modeling of transmission networks: an explanatory study on further EHV underground cabling in the netherlands, in *2016 IEEE International Energy Conference (ENERGYCON)* (2016), pp. 1–6

45. B.W. Tuinema, Reliability of transmission networks - impact of EHV underground cables and interaction of offshore-onshore networks. Ph.D. dissertation, Delft University of Technology (2017), https://repository.tudelft.nl/

46. A.H.M.E. Reinders, M. de Respinis, J. van Loon, W. Schram, W. van Sark, E. Gultekin, B. van Mierlo, C. Robledo, I. Papaioannou, A. van Wijk, A. Stekelenburg, F. Bliek, T. Esterl, S. Uebermasser, F. Lehfuss, M. Lagler, E. Schmautzer, T. Hahn, L. Fickert, Co-evolution of smart energy products and services: a novel approach towards smart grids, in *Asian Conference on Energy, Power and Transportation Electrification (ACEPT)* (Singapore, 2016), pp. 25–27, Oral presentation: http://ieeexplore.ieee.org/document/7811522/

47. C.B. Robledo, V. Oldenbroek, F. Abbruzzese, A. van Wijk, Integrating a hydrogen fuel cell electric vehicle with vehicle-to-grid technology, photovoltaic power and a residential building. Appl. Energy **215**, 615–629 (2017). https://doi.org/10.1016/j.apenergy.2018.02.038

48. S.H. Strogatz, Exploring complex networks. Nature **410**, 268 (2001)

49. H. Cetinay, S. Soltan, F.A. Kuipers, G. Zussman, P. Van Mieghem, Comparing the effects of failures in power grids under the AC and DC power flow models. IEEE Trans. Netw. Sci. Eng. (2017)

50. H. Cetinay, F.A. Kuipers, P. Van Mieghem, A topological investigation of power flow. IEEE Syst. J. (2016)

51. H. Cetinay, T. Kekec, F.A. Kuipers, D.M.J. Tax, Markov random field for wind farm planning, in *The 5th IEEE International Conference on Smart Energy Grid Engineering* (2017)

52. H. Cetinay, F.A. Kuipers, A.N. Guven, Optimal siting and sizing of wind farms. Elsevier Renew. Energy **101**, 51–58 (2017)

53. R. Künneke, D.C. Mehos, R. Hillerbrand, K. Hemmes, Understanding values embedded in offshore wind energy systems: toward a purposeful institutional and technological design. Environ. Sci. Policy **53**, 118–129 (2015)

54. C. Milchram, R. Hillerbrand, Values at stake for smart grid systems: the case of the Netherlands, in *Proceedings of the 12th Conference on Sustainable Development of Energy, Water, and Environment Systems* (2017)

55. T.E. De Wildt, E.J.L. Chappin, G. Van De Kaa, P. Herder, A comprehensive approach to reviewing latent topics addressed by literature across multiple disciplines. Appl. Energy (in press)

56. P. Ayivor, J. Torres, M.A.M.M. van der Meijden, R. van der Pluijm, B. Stouwie, Modelling of large size electrolyzer for electrical grid stability studies in real time digital simulation, in *Proceedings of Energynautics 3rd International Hybrid Power Systems Workshop* (Tenerife, Spain, 2018)

57. G. Suárez, P. Ayivor, J.R. Torres, M.A.M.M. van der Meijden, Demand side response in multi-energy sustainable systems to support power system stability, in *Proceedings of Energynautics 16th Wind Integration Workshop* (Springer, Berlin, 2017)

58. V.G. Suárez, J.L.R. Torres, B.W. Tuinema, A.P. Guerra, M.A.M.M. van der Meijden, Integration of Power-to-gas conversion into dutch electrical ancillary services markets, in *Proceedings of ENERDAY 2018 - 12th International Conference on Energy Economics and Technology* (Dresden, Germany, 2018)

59. TSO2020 Consortium, TSO2020 (Electric "Transmission and Storage Options" along TEN-E and TEN-T corridors for 2020) project website (2018), www.tso2020.eu
60. T.I. Strasser, et al., An Integrated Research Infrastructure for Validating Cyber-Physical Energy Systems. in *Industrial Applications of Holonic and Multi-Agent Systems*, ed. by v. Mařík, W. Wahlster, T. Strasser, P. Kadera. Lecture Notes in Computer Science, vol. 10444 (Springer, cham, Berlin, 2017) HoloMAS
61. V.H. Nguyen et al., Power-hardware-in-the-loop experiments together with co-simulation for the holistic validation of cyber-physical energy systems, in *IEEE PES Innovative Smart Grid Technologies Conference Europe (ISGT-Europe)* (Torino, 2017), pp. 1–6
62. A.A. van der Meer et al., Cyber-physical energy systems modeling, test specification, and co-simulation based testing, in *Workshop on Modeling and Simulation of Cyber-Physical Energy Systems (MSCPES)* (Pittsburgh, PA, 2017) pp. 1–9

Part II
Methods for Analysis of Future Power Grids

Chapter 2
Topology-Driven Performance Analysis of Power Grids

Hale Çetinay, Yakup Koç, Fernando A. Kuipers and Piet Van Mieghem

Abstract Direct connections between nodes usually result in efficient transmission in networks. Such electric power transmission is governed by physical laws, and an assessment purely based on direct connections between nodes and shortest paths may not capture the operation of power grids. Motivated by these facts, in this chapter, we investigate the relation between the electric power transmission in a power grid and its underlying topology. Initially, we focus on synthetic power grids whose underlying topology can be structured as either a path or a complete graph. We analytically compute the impact of electric power transmission on link flows under the normal operation and under a link failure contingency using the linearised DC power flow equations. Subsequently, in various other graph types, we provide empirical results on the link flow, the voltage magnitude and the total active power loss in power grids using the nonlinear AC power flow equations. Our results show that in a path graph, as an assessment based on shortest paths holds, however, the electric power transmission can lead to substantial amount of link flows, active power loss and voltage drops, especially in large path graphs. On the other hand, adding few links to a path graph could significantly improve those performance indicators of power grids, but at a cost: the resulting meshed topology decreases the control over power grids

H. Çetinay (✉) · F. A. Kuipers · P. Van Mieghem
Faculty of Electrical Engineering, Mathematics and Computer Science, P.O Box 5031, 2600 GA Delft, The Netherlands
e-mail: H.Cetinay-Iyicil@tudelft.nl

F. A. Kuipers
e-mail: F.A.Kuipers@tudelft.nl

P. Van Mieghem
e-mail: P.F.A.VanMieghem@tudelft.nl

Y. Koç
Risk Management Group, Asset Management, Stedin B.V., 2628 CD Delft, The Netherlands
e-mail: yakupkoc@gmail.com

Y. Koç
Systems Engineering, Technology Policy and Management, Delft University of Technology, 2628 CD Delft, The Netherlands

© Springer Nature Switzerland AG 2019
P. Palensky et al. (eds.), *Intelligent Integrated Energy Systems*,
https://doi.org/10.1007/978-3-030-00057-8_2

as a direct assessment between the shortest paths and the electric power transformation is lost. Additionally, a meshed topology with loops increases the redundancy in the design to ensure a safe operation under a link failure contingency.

2.1 Introduction

Many researchers analyse power grids from a graph topological point of view [1, 2]. Various topology metrics (such as nodal degree, clustering coefficient) have been proposed to assess the vulnerability or to locate the critical components of power grids [3–5]. Those purely topological approaches, however, may fail to fully capture the physical and operational specific features of power grids whose operation are governed by physical laws.

In this chapter, we take an extended graph theoretical approach [6–8] by modelling the electrical properties such as flow allocation according to Kirchhoff's laws and the impedance values of transmission lines in power grids. We investigate the impact of network topology on the key performance indicators of power grids, which we take as the node voltage, the link flow, the total power loss and the served electric demand.

Initially, we focus on the operation by considering two extreme graphs. In synthetic power grids whose underlying topology is either a path or a complete (full-mesh) graph, we analytically derive the steady-state operating conditions under normal operation and under a random link failure (removal) contingency using the linearised DC power flow equations. Subsequently, in various other graphs, we empirically investigate the relation between the topology and the key performance indicators using the nonlinear AC power flow equations [9].

The remainder of this chapter is organized as follows. Section 2.2 investigates the electric power transmission in path and complete graphs under normal operation. In Sect. 2.3, we focus on single link failure contingencies in those graphs, and derive the impact of a random link failure on the steady-state link flows. Section 2.4 presents our empirical results on the key performance indicators in various graphs both under normal operation and single link failure contingencies. Section 2.5 concludes the chapter.

2.2 DC Power Flow Analysis in Path and Complete Graphs

A connected simple graph (i.e., a graph with no parallel duplicate links or self-loops) lies between a complete graph and a tree graph. In a complete graph, every pair of distinct nodes is connected by a link. On the other hand, a tree has no cycles; consequently, any two nodes are connected by exactly one path.

The direct connections between nodes usually result in an efficient transmission in a network. The distances between the nodes in a complete graph are shortest compared to the other graphs, in which multiple hops are needed to reach the destination. In power grids, different than the typical transmissions based on the shortest paths, the electric power transmission is governed by physical laws. Therefore, an assessment based on purely the direct connections between nodes may not be enough to draw conclusions. In this section, we investigate the electric power transmission in those extreme graph types.

We model power grids with N buses (nodes), and L lines (links) by a weighted graph $G(N, L)$. We use $\mathcal{N}$ to denote the set of N nodes and $\mathcal{L}$ to denote the set of L links with equal weights, b. Every link $l_{ik} \in \mathcal{L}$ is associated with a maximum flow capacity C_{ik} that represents the maximum power flow that can be afforded by the corresponding line, and a rest flow capacity $\alpha_{ik} = C_{ik} - |f_{ik}|$ where $|f_{ik}|$ is the flow through the link l_{ik} under normal operation. We assume a single upstream *supply* node, and treat the remaining $N - 1$ downstream nodes as *demand* nodes. Without loss of generality, we label the supply node as node 1, and take the total electric power demand of the network as $(N - 1)p$ where $p \geq 0$ is a constant. Throughout Sects. 2.2 and 2.3, we adopt the slack-bus independent solution to the DC power flow equations [10], which could approximate the steady-state operation under the DC power flow assumptions [11].

2.2.1 Electric Power Transmission in a Path Graph

We investigate the electric power transmission from the supply node 1 to the single demand node N in a path graph (whose nodes are labeled consecutively). The magnitude $|f_{ik}^{1 \to N}|$ of the flow through a link l_{ik} between node i and node $k = i + 1$ is found as (see Sect. 2.6.1)

$$|f_{ik}^{1 \to N}| = b|(\theta_i - \theta_k)|$$
$$= p(N - 1) \quad \forall l_{ik} \in \mathcal{L}, \tag{2.1}$$

where b is the reciprocal of the line reactance and θ_i is the phase angle of the voltage at node i.

Equation (2.1) shows that the resulting link flows due to the electric power transmission are all the same, and their values increase with the increasing graph size N and unit power demand p. This linear correlation between the size and the magnitudes of link flows could lead to substantial flows and result in congestion problems, especially in large graphs.

On the other hand, in a path graph, the electric power is transferred through a single path between the supply and the demand node. Consequently, an assessment based on shortest paths holds, which can ease the supervision of the network operator.

2.2.2   Electric Power Transmission in a Complete Graph

Similar to Sect. 2.2.1, we investigate the electric power transmission from the supply node 1 to a single (randomly chosen) demand node m in a complete graph. The magnitude $|f_{ik}^{1 \to m}|$ of the resulting flow through a link l_{ik} is found as (see Sect. 2.6.2)

$$|f_{ik}^{1 \to m}| = \begin{cases} \frac{2(N-1)p}{N} & \text{if } l_{ik} = l_{1m}, \\ \frac{(N-1)p}{N} & \text{if } l_{ik} \in \left\{ \left(\mathcal{B}(1) \cup \mathcal{B}(m) \right) \setminus l_{1m} \right\}, \\ 0 & \text{otherwise}, \end{cases} \tag{2.2}$$

where $\mathcal{B}(i)$ denotes the direct neighbors of node i.

Equation (2.2) indicates that three different magnitudes of link flow exist during the electric power transmission: (a) The flow through the link between the supply and the demand node is maximum, whereas (b) the flows through the links to the other neighbors of those nodes are half of that maximum flow, and (c) the remaining links that are not direct neighbors of either the supply or the demand node have zero flows.

Comparing the magnitudes of link flow in a path graph in Eq. (2.1) and a complete graph in Eq. (2.2) shows that the maximum link flow due to the electric power transmission from the supply node to a demand node is dramatically lower in a complete graph. However, the distribution of the flows through links in a complete graph is not homogeneous, thus the relation between the total decrease in the magnitudes of link flow and the total number of links added to a path graph is not linear.

2.3   DC Power Flow Analysis in Path and Complete Graphs After a Random Link Failure

Single line failures are common in power grids. Therefore, as well as under the normal operation, the operation after a link failure (removal) is important to assess the reliability of power grids [12]. In this section, we theoretically investigate the effect of a random link failure on link flows in path and complete graphs using the linearised DC power flow equations. In addition, to quantify the effect of link failure contingencies in a graph, we calculate the *theoretical robustness function* of those graphs, which we define as the expected fraction of served demands after a random link failure.

2.3.1   Random Link Failure in a Path Graph

First, we focus on the effect of a single link failure in path graphs. Just before the link failure takes place, we assume that all demand nodes have a unit electric power demand of p, which we refer to as the *symmetrical distribution* of demands. In other words, the supply node transfers a unit electric power of p to every other demand node, resulting in the magnitude $|f_{ik}|$ of the flow through link l_{ik} between node i and node $k = i + 1$ before the failure

$$|f_{ik}| = p(N - i). \tag{2.3}$$

As the graph under investigation is a path graph with no cycles, the removal of any link l_{ik} between node i and node $k = i + 1$ partitions the graph (see Sect. 2.6.3), and this partition removes in total $p \times (N - i)$ demand from the graph according to Eq. (2.3). Therefore, the closer the link failure is to the supply node 1, the worse is the effect on the served demands.

The continuity of the operation of the network depends on the location of the failed link. If the failed link is adjacent to the supply node, then the supply node is isolated from the demand nodes and the network faces a complete blackout. If the failure probabilities of the links are the same in a path graph, the probability p_b that a random link failure leads to a complete blackout is $p_b = \frac{1}{L} = \frac{1}{N-1}$.

The failure and removal of any other link partitions the network and the remaining network can continue functioning, though with decreased demands. As the total demand of the network decreases after the link removal, the flows through the remaining links decrease, and thus, there is no possibility for further cascading failures [13] due to the insufficient rest flow capacity of the remaining links. Consequently, the expected fraction $E[F_s]$ of served demands after a random link failure is

$$E[F_s] = \frac{1}{N-1} \times \frac{(0 + 1 + \cdots + N - 2)}{N - 1} = \frac{N - 2}{2(N - 1)}.$$

2.3.2 Random Link Failure in a Complete Graph

Next, we investigate a random link failure in a complete graph. After the removal of a link, the flows are redistributed following Kirchhoff's laws and the flows through the remaining links may change. Due to the meshed topology of the graph, this redistribution can lead to an increase or a decrease in flow through a particular link [10].

Before a link failure happens, under symmetrical distribution of demands, i.e., when the supply node transfers a unit demand of p to every other demand node, the magnitude $|f_{ik}|$ of the flow through a link l_{ik} in a complete graph is

$$|f_{ik}| = \begin{cases} p & \text{if } l_{ik} \in \mathcal{B}(1), \\ 0 & \text{otherwise.} \end{cases} \tag{2.4}$$

Following Eq. (2.4), two different magnitudes of link flows exist in a complete graph under the symmetrical distribution of demands. As a result, a single link failure and removal can result in two cases:

Failure of a Link with Zero Flow

When a link l_{ik} is removed from the graph, the flow $|f_{ik}|$ through the link before failure needs to be redistributed over the alternative paths between nodes i and k. Since a *redundant* link l_{ik} does not transport any flow, its removal does not cause a power redistribution.

Failure of a Link with Maximum Flow

When a *used* link l_{ik} is removed from the graph, the flow $|f_{ik}| = p$ through the link before failure is redistributed over alternative paths between nodes i and k. As a result, the initial link l_{ik} failure may trigger further failures in the network if the increase $|\Delta f_{ab}| = \frac{p}{N-2}$ in the flow through a remaining link l_{ab} exceeds its rest flow capacity α_{ab} (see Sect. 2.6.4),

$$|\Delta f_{ab}| = \frac{p}{N-2} \geq \alpha_{ab}. \tag{2.5}$$

When the rest flow capacity α_{ab} is smaller than the required value in Eq. (2.5), consecutive failures occur. After the initial failure of the used link l_{ik}, the flows through all remaining used links exceed their maximum flow capacity, and fail in the next stage of the failure. This isolates the supply node. Consequently, the remaining network cannot match any demands, and it could face a complete blackout.

When the size N of the graph is 2, i.e., when there is only one link, the failure of that link destroys the graph by separating the supply node from the demand node regardless of the rest flow capacity α_{ab} of the link. For larger graphs, Eq. (2.5) shows the required rest flow capacity α_{ab} of links is maximum when the size of the graph is $N = 3$, whereas it decreases as N increases. This means the effect of a link removal on the flows through the remaining links reduces with the size N of the graph.

Finally, we calculate the *theoretical robustness function* of a complete graph, which is the expected fraction $E[F_s]$ of served demands after a random link removal from the underlying graph. Figure 2.1 presents the theoretical robustness function of a complete graph under the symmetrical distribution of demands. If the rest flow capacity of the links is larger than the required value in Eq. (2.5), the remaining links can tolerate the redistributed flows after a random link removal. The network can continue to serve the same amount of total demand after any single link failure. Therefore, in region II in Fig. 2.1, the fraction of served demands stays the same. On the other hand, when the rest flow capacity of the links is smaller than the required value in Eq. (2.5) in region I, the network continues its operation only if the failed link is a *redundant* link with zero flow. Otherwise, when a *used* link with flow p fails, the network faces complete blackout and cannot serve any demand. Therefore, the expected $E[F_s]$ fraction of served demands after a random link failure in region I is calculated as

Fig. 2.1 The expected $E[F_s]$ fraction of served demands versus the rest flow capacity of the links after a random link failure in a complete graph under symmetrical distribution of demands. The figure is computed for $N = 5$

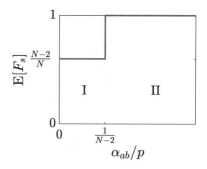

$$E[F_s] = 1 \times p_r + 0 \times p_u = \frac{N - 2}{N} = 1 - \frac{2}{N},$$

where $p_r = \frac{N-2}{N}$ represents the probability that the failed link is a redundant link with zero flow and $p_u = \frac{2}{N}$ represents the probability that the failed link is a used link with flow p (see Eqs. (2.16) and (2.17)).

Comparing the effect of single link failures in a path and a complete graph shows that a path (or a tree) cannot provide a back-up path after a link removal, and the total served demand in the network definitely decreases. In addition, as the demands in the network decrease, there is no possibility of cascading failures in a tree graph with no loops. On the other hand, a meshed topology can provide back-up paths during random link removals. Yet, the correct setting of design parameters, i.e., the rest flow capacities of the links, is extremely important. If the rest flow capacity α_{ab} of the links is smaller than in Eq. (2.5) and a used link fails, then the remaining links in a complete graph cannot tolerate the redistributed flows. Consequently, a random link removal in a complete graph could lead to more link and demand losses than a link failure in a simple path graph.

2.4 Impact of Topology on Key Performance Indicators of Power Grids

In Sects. 2.2 and 2.3, we focus on the theoretical analyses of electric power transmission in path and complete graphs. In this section, we focus on many other graphs and empirically investigate the key performance indicators of power grids under normal operation as well as under single link failure contingencies. In our analyses, we use the AC power flow solver in Matlab [9] to calculate the steady-state operating conditions under the symmetrical distribution of demands. In addition to the line reactance x and the active power p values, we take into account the line resistance r and the reactive power q values for a more practical model of power grids.

2.4.1 Key Performance Indicators Under Normal Operation

For the safe and efficient operation of power grids, lower magnitudes of link flow and total power loss, and higher values of node voltage (close to 1 per unit) are desired.[1] In a power grid whose topology are modelled by the specific graph G, we define the satisfaction degree of performance indicators of link flow $\zeta(G)$, node voltage $v(G)$ and power loss $\eta(G)$ as

$$\zeta(G) = \begin{cases} 1 & \text{if } \max_{l_{ik} \in \mathcal{L}(G)} \left(|f_{ik}|\right) < p, \\ \frac{\tau_f - \max_{l_{ik} \in \mathcal{L}(G)} \left(|f_{ik}|\right)}{\tau_f - p} & \text{if } p \leq \max_{l_{ik} \in \mathcal{L}(G)} \left(|f_{ik}|\right) \leq \tau_f, \\ 0 & \text{if } \max_{l_{ik} \in \mathcal{L}(G)} \left(|f_{ik}|\right) > \tau_f, \end{cases} \qquad (2.6)$$

$$v(G) = \begin{cases} 0 & \text{if } \min_{i \in \mathcal{N}(G)} \left(v_i\right) < \tau_v, \\ \frac{\min_{i \in \mathcal{N}(G)} \left(v_i\right) - \tau_v}{1 - \tau_v} & \text{if } \tau_v \leq \min_{i \in \mathcal{N}(G)} \left(v_i\right) \leq 1, \\ 1 & \text{if } \min_{i \in \mathcal{N}(G)} \left(v_i\right) > 1, \end{cases} \qquad (2.7)$$

$$\eta(G) = \begin{cases} 1 & \text{if } \tau_\sigma < 0, \\ \frac{\tau_\sigma - \sigma(G)}{\tau_\sigma} \leq \tau_f, & \text{if } 0 \leq \sigma(G) \leq \tau_\sigma, \\ 0 & \text{if } \sigma(G) > \tau_\sigma, \end{cases} \qquad (2.8)$$

where $|f_{ik}|$ is the magnitude of the flow through link l_{ik}, v_i is the magnitude of voltage at node i, $\sigma(G)$ is the total active power loss, and $\mathcal{L}(G)$ and $\mathcal{N}(G)$ denote the set of links and nodes of graph G, respectively. The performances in Eqs. (2.6)–(2.8) are evaluated on a scale from 0 to 1 (see Fig. 2.2): The highest performance of 1 corresponds to *ideal power grids* in which the maximum link flow is equal to the unit power demand p, the minimum voltage is equal to 1 per unit, i.e., no voltage drop, and the total power loss is 0, i.e., a lossless power grid. Conversely, the lowest performance of 0 corresponds to the maximum link flow τ_f, the minimum node voltage τ_v and the total power loss τ_σ. The requirements of τ_f, τ_v and τ_σ are usually determined by the specific grid codes of the operators.

Figure 2.3 shows the variations of the key performance indicators under the symmetrical demand distribution throughout the topological transformation of the path graph with 5 nodes and 4 links.[2] Figure 2.3 depicts that the performance indicator of link flow is lowest in the path, and highest in the complete graph. We observe that the cycle graph *dramatically increases the performance indicator of link flow* by decreasing the maximum link flow compared to the path graph. Similar to the

---

[1] As we focus on the impact of electric power transmission from a supply node to demand nodes, only voltage drops in the network are considered.

[2] We compute all possible ways to evaluate the transformation from the path to the complete graph, which is only possible for small graphs.

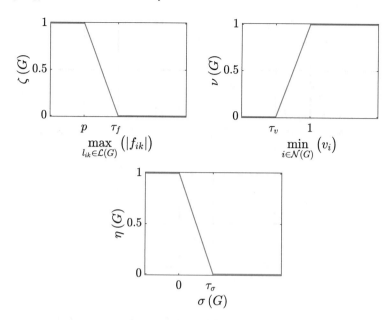

Fig. 2.2 The functions of the performance indicators of link flow $\zeta(G)$, node voltage $\nu(G)$ and power loss $\eta(G)$

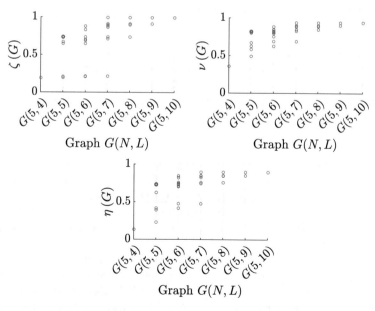

Fig. 2.3 The variations of the key performance indicators throughout the topological transformation of the path graph $G(5, 4)$ with 5 nodes and 4 links. The transformation towards the complete graph $G(5, 10)$ requires the addition of $\frac{N(N-1)}{2} - (N - 1) = 6$ links, which can be performed in $2^6 - 1 = 63$ different ways. The bold red data point corresponds to the cycle graph. The performance indicators are evaluated for $r = 0.1$, $x = 0.1$, $p = 0.05$, $q = 0.01$, $\tau_f = 5p$, $\tau_v = 0.9$ and $\tau_\sigma = 0.2p$

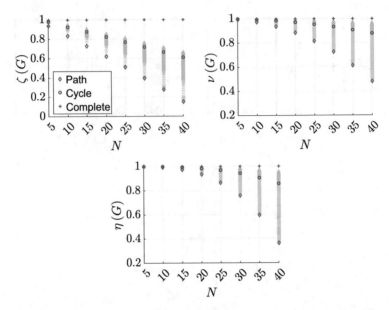

Fig. 2.4 The variations of the key performance indicators in graphs with different sizes N. The blue lower triangle (∇), and green upper triangle ($\triangle$) data points correspond to the graphs that are constructed by adding one link to the path graph, and adding one link to the cycle graph, respectively. The performance indicators are evaluated for $r = 0.02, x = 0.02, p = 0.005, q = 0.001, \tau_f = 50p$, $\tau_v = 0.8$ and $\tau_\sigma = 4p$

link flows, the minimum voltage is lowest and the total power loss is highest in the path graph. The complete graph, on the other hand, represents the operation at the minimum voltage drop and total power losses, thus corresponding to the highest values of key performance indicators.

Figure 2.3 indicates that a meshed topology can improve the key performance indicators compared to the path graph. We showed in Eq. (2.2) that the flow distribution in a complete graph due to an electric power transmission is not homogeneous, which could explain the nonlinear relation between the total number of added links to the path graph and the total increase in the key performance indicators in Fig. 2.3. In particular, the cycle graph and the *augmented cycles*, i.e., the graphs constructed from the cycle graph by adding links, are observed to affect the key performance indicators dramatically.

In Fig. 2.4, we present the variations of key performance indicators in the graphs with different sizes N. In the complete graphs, the maximum link flows are nearly the same for all sizes N, which is in agreement with the theoretical calculations in Eq. (2.4). In the other graphs, the maximum link flow increases with increasing size N, decreasing the performance indicator of link flow.

From Fig. 2.4, we observe that the minimum values of node voltages are nearly the same in the complete graphs, whereas they decrease dramatically in the path graphs with increasing size N. On the other hand, the cycle topology increases the node voltages, thus also the performance indicator of node voltage, rapidly compared to the path graph.

Similar observations hold for the performance indicator of power loss. In large path graphs, the total active power loss of the network is high, which decreases the performance indicator of power loss. On the other hand, a complete graph nearly zeroes the power loss and the cycle topology significantly decreases the total loss compared to a path graph.

Similar to Figs. 2.3 and 2.4 illustrates that a meshed topology can improve the key performance indicators compared to a path graph. We conclude that the core contributions to the key performance indicators arise from the first few links added to a path graph. In particular, for larger graphs, a cycle topology can dramatically increase the voltage magnitude and decrease total active power loss of the network compared to the path graph. Consequently, adding a limited number of links to the tree topology can still achieve higher levels of performance during the electric power transmission between a supply and demand nodes.

2.4.2 Key Performance Indicators Under a Single Link Failure Contingency

In this section, we investigate the effect of a single link failure in different graphs. Initially, we focus on the effect of a link failure on the served demands of the network. Figure 2.5 illustrates the expected $E[F_s]$ fraction of served demands after a random link failure throughout the topological transformation of a path graph $G(5, 4)$ with 5 nodes and 4 links. In Sect. 2.3, we show that any link removal from a path graph partitions the graph. Figure 2.5 also depicts that only a cycle or augmented cycles can provide a back-up after any random link failure. The other graphs may partition after a random link failure and can continue their operation only with a decreased

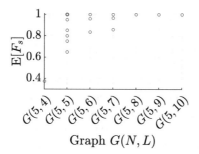

Graph $G(N, L)$

Fig. 2.5 The expected $E[F_s]$ fraction of served demands after a random link failure throughout the topological transformation of a path graph with 5 nodes. The bold red data point corresponds to the cycle graph. The remaining links are assumed to have enough rest flow capacity to handle the redistributed flows due to a random link failure

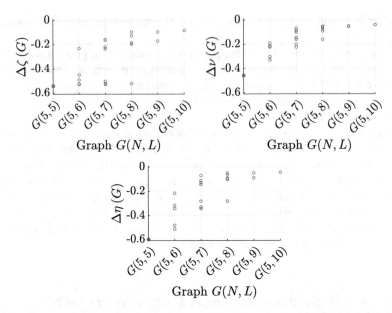

Fig. 2.6 The variations of the changes Δ in the key performance indicators after a link failure throughout the topological transformation of a path graph with 5 nodes. The bold red data point corresponds to the cycle graph. The performance indicators are evaluated for $r = 0.1$, $x = 0.1$, $p = 0.05$, $q = 0.01$, $\tau_f = 5p$, $\tau_v = 0.9$ and $\tau_\sigma = 0.2p$

total demand, which usually improves the key performance indicators. Therefore, in this subsection, we only focus on the graphs that can provide a back-up after any random link failure.

To investigate and compare the effect of single link failures in each graph, we removed one link at a time from the graph, and calculated the changes in the performance indicators. We repeated this link failure contingency simulation for each link, and compared all changes in the link flow, the node voltage and the active power loss in the network. Figure 2.6 illustrates the maximum resulting changes in these key performance indicators after a link removal throughout the topological transformation of a path graph $G(5, 4)$ with 5 nodes and 4 links. The performance indicator of link flow can significantly decrease after a link failure in a cycle graph. For the complete graph, on the other hand, we observe that the effect of a link failure on the indicator of link flow is very small. Similar to the changes in the indicator of link flow, the decreases in the performance indicators of node voltage and power loss are highest in the cycle graph after a single link failure.

Finally, in Fig. 2.7, we present the variations of key performance indicators after a link failure in graphs with different sizes N. Similar to the theoretical calculation in Eq. (2.5), the effect of a link failure on the remaining link flows slightly decreases with the increasing size N in the complete graphs. On the other hand, in cycle graphs, the change in the magnitude of the flow through a remaining link can significantly increase with the increasing size N, which decreases the related performance indi-

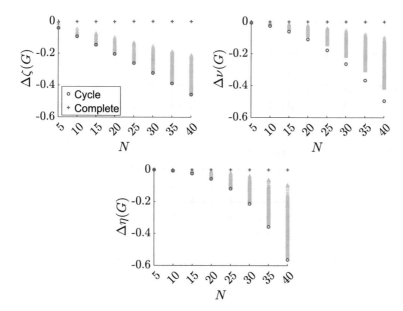

Fig. 2.7 The variations of the changes Δ in the key performance indicators after a link failure in graphs with different sizes N. The green upper triangle ($\triangle$) data points correspond to the graphs that are constructed by adding one link to the cycle graph. The performance indicators are evaluated for $r = 0.02$, $x = 0.02$, $p = 0.005$, $q = 0.001$, $\tau_f = 50p$, $\tau_v = 0.8$ and $\tau_\sigma = 4p$

cator. In the *worst* case, when one of the links adjacent to the supply node fails in a cycle graph, it operates as a path graph with the same size N after the link failure. Therefore, the performance indicator of link flow in a cycle graph under a single link failure contingency becomes the performance indicator of link flow in a path graph under normal operation.

Similar to the changes in the performance indicator of link flow after a link failure, the indicators of node voltage and power loss slightly decrease in a complete graph with increasing size N. In the other graphs, however, the decrease in the key performance indicators could be drastic. Although under the normal operation, the cycle and the augmented cycles can provide higher values of the performance indicators; after a link failure, large drops on the key performance indicators in those graphs are expected.

2.5 Conclusion

In this chapter, we investigated the impact of topology on the electric power transmission and the performance of power grids. By utilizing a graph theoretical approach, we first focused on two extreme graphs, complete and path graphs, and analysed the electric power transmission under normal operation and under single link failure contingencies. We showed that in complete graphs, due to the redistributed flows, the survival of power grids from a random link failure depends on the rest flow capacity of the remaining links. Consequently, when the rest flow capacities are insufficient to handle the redistributed flows, a single link failure could result in more link and demand loss in a complete graph than in a path graph.

Subsequently, we empirically investigated the effect of the electric power transmission on the link flow, the node voltage and the active power loss in power grids in various other graphs. Our results show that adding few links to a path graph can significantly improve these key performance indicators of power grids compared to a path graph. However, at the same time, the performance indicators could also remarkably decrease after a link failure. Consequently, throughout a topological transformation towards a meshed topology with loops, redundancies in the design parameters of power grids are needed to ensure safety under normal operation and as well as under single link failure contingencies.

Acknowledgements This work was supported in part by Alliander N.V.

Appendix

2.6 DC Power Flow Equations in Extreme Graph Types

We model power grids with N buses (nodes), and L transmission lines (links) by a weighted graph $G(N, L)$. The $N \times N$ weighted adjacency matrix $\mathbf{W}$ specifies the interconnection pattern of the graph $G(N, L)$: w_{ik} is non-zero only if the nodes i and k are connected by a link; otherwise $w_{ik} = 0$. In the slack-bus independent solution of the DC power flow equations [10], the relation between the phase angles $\Theta = [\theta_1 \ldots \theta_N]^T$ of node voltages, and the power input $\mathbf{P} = [p_1 \ldots p_N]^T$ is given as [10]

$$\Theta = \mathbf{Q}^+\mathbf{P}, \tag{2.9}$$

where $\mathbf{Q}^+$ is the pseudo inverse of the Laplacian $\mathbf{Q}$ of the weighted graph $G(N, L)$.

2.6.1 Operating Conditions in a Path Graph

For a path graph, whose nodes are numbered consecutively and links have equal link weights $b > 0$, the structure of weighted Laplacian $\mathbf{Q}$ can be written as

$$Q = \begin{bmatrix} b & -b & 0 & \dots & 0 & 0 \\ -b & 2b & -b & 0 & \dots & 0 \\ 0 & \ddots & \ddots & \ddots & & \vdots \\ \vdots & & \ddots & \ddots & \ddots & 0 \\ 0 & \dots & 0 & -b & 2b & -b \\ 0 & 0 & \dots & 0 & -b & b \end{bmatrix}.$$

In order to find Q^+ in Eq. (2.9), we use the definition [14] of the pseudo-inverse of Laplacian $Q^+ = \hat{X} \mathbf{diag}\left(\frac{1}{\mu_k}\right) \hat{X}^T$, where the $N \times (N-1)$ matrix $\hat{X}$ consists of all the normalized eigenvectors of Q, except for the eigenvector u belonging to eigenvalue $\mu = 0$, and where the $(N-1) \times (N-1)$ diagonal matrix $\mathbf{diag}\left(\frac{1}{\mu_k}\right)$ contains the positive eigenvalues of Laplacian Q.

The positive eigenvalues μ_k and the corresponding normalized eigenvector elements $\hat{X}(v, k)$ of the weighted Laplacian of a path graph are [8]

$$\mu_k = 2b \left(1 - \cos\left(\frac{\pi k}{N}\right)\right),$$

$$\hat{X}(v, k) = \frac{\sqrt{2}}{\sqrt{N}} \times \cos\left(\frac{\pi k v}{N} - \frac{\pi k}{2N}\right),$$

where $1 \leq k \leq N - 1$, and $1 \leq v \leq N$. Then, the elements q_{ik}^+ of the pseudo-inverse of the Laplacian are

$$q_{ik}^+ = \sum_{v=1}^{N-1} \frac{\hat{X}(i, v)\hat{X}(k, v)}{\mu_v}. \tag{2.10}$$

Inserting the elements of pseudo-inverse in Eq. (2.10) and the power input $P = [(N-1)p, 0, \dots, 0, -(N-1)p]^T$ into the DC power flow equations in Eq. (2.9) results in the operating conditions, i.e., the phase angles Θ of node voltages, when the electric power is transferred from the supply node 1 to the demand node N:

$$\theta_i = p \times (N-1) \times (q_{i1}^+ - q_{iN}^+)$$
$$= \frac{p(N-1)(N-2i+1)}{2b}. \tag{2.11}$$

2.6.2 Operating Conditions in a Complete Graph

For a complete graph with equal link weights $b > 0$, the structure of the weighted Laplacian Q can be written as

$$\mathbf{Q} = b(N\mathbf{I} - \mathbf{J}), \tag{2.12}$$

where $\mathbf{J}$ is the all-one matrix, and $\mathbf{I}$ is the identity matrix. Using the definition of pseudo-inverse of the Laplacian [14]

$$\mathbf{Q}^+ = (\mathbf{Q} + \alpha\mathbf{J})^{-1}\left(\mathbf{I} - \frac{1}{N}\mathbf{J}\right), \tag{2.13}$$

where $\alpha > 0$ is a scalar, and choosing the scalar $\alpha = b$, the pseudo-inverse of the weighted Laplacian of a complete graph can be found as

$$\mathbf{Q}^+ = (\mathbf{Q} + b\mathbf{J})^{-1}\left(\mathbf{I} - \frac{1}{N}\mathbf{J}\right)$$
$$= \frac{1}{Nb}\left(\mathbf{I} - \frac{1}{N}\mathbf{J}\right) = \frac{1}{N^2 b}(N\mathbf{I} - \mathbf{J}). \tag{2.14}$$

From Eq. (2.9), the phase angles Θ of node voltages when the electric power transferred from the supply node 1 to the demand node N can be found as

$$\Theta = \mathbf{Q}^+\mathbf{P}$$
$$= \frac{1}{bN^2}\begin{bmatrix} (N-1) & -1 & \cdots & & -1 \\ -1 & \ddots & & & \vdots \\ \vdots & & \ddots & & -1 \\ -1 & \cdots & & -1 & (N-1) \end{bmatrix}\begin{bmatrix} (N-1)p \\ 0 \\ \vdots \\ -(N-1)p \end{bmatrix} = \frac{p}{bN}\begin{bmatrix} (N-1) \\ 0 \\ \vdots \\ -(N-1) \end{bmatrix}.$$

2.6.3 Single Link Failure in a Path Graph

The failure and removal of a link l_{ik} from a network partitions its underlying graph if the equality between the reactance x_{ik} of the link and the effective resistance r_{ik} between its node pairs satisfies [10]

$$x_{ik} = r_{ik}.$$

When the underlying topology is a path graph, the effective resistance r_{ik} between nodes i and $k = i + 1$ can be written as

$$r_{ik} = |i - k| \times x_{ik},$$

meaning that the removal of any link l_{ik} partitions the path graph.

2.6.4   Single Link Failure in a Complete Graph

When a link l_{ik} with flow $|f_{ik}| = p$ is removed from the graph, the flow $|f_{ik}| = p$ through the link before its removal is redistributed over alternative paths between nodes i and k. Hence, the final flow through an arbitrary remaining link l_{ab} can be written as the sum of the previous state of the network, i.e., the previous flow through the link between nodes a and b when link l_{ik} is present, and the flow resulting from the change of the state due to the removal of link l_{ik}. The change in the flow Δf_{ab} through a remaining link l_{ab} can be calculated as [10]

$$\Delta f_{ab} = w_{ab} \times \frac{(r_{ak} - r_{ai} + r_{bi} - r_{bk})}{2(1 - w_{ik} \times r_{ik})} \times p, \tag{2.15}$$

where w_{ab} is the weight of the link l_{ab} and r_{ak} is the effective resistance between the nodes a and k.

The effective resistance between any two distinct nodes in the complete graph with equal link weights b is $\frac{2}{bN}$. Therefore, the numerator $(r_{ak} - r_{ai} + r_{bi} - r_{bk})$ of Eq. (2.15) is nonzero and its magnitude is equal to $|r_{ak} - r_{ai} + r_{bi} - r_{bk}| = \frac{2}{bN}$ only when the removed link l_{ik} and the observed link l_{ab} share a node. Then,

$$|\Delta f_{ab}| = \begin{cases} 0 & \text{if } l_{ik} \cap l_{ab} = \varnothing, \\ \frac{p}{N-2} & \text{otherwise.} \end{cases}$$

If the failure probabilities of the links in a complete graph are the same, we can calculate the probability p_r that a failed link is a *redundant* link with zero flow, and the probability p_u that a failed link is a *used* link with non-zero flow as

$$p_r = \frac{(N-1)(N-2)}{2} \times \frac{2}{N(N-1)} = \frac{N-2}{N} = 1 - \frac{2}{N}, \tag{2.16}$$

$$p_u = (N-1) \times \frac{2}{N(N-1)} = \frac{2}{N}, \tag{2.17}$$

where $p_u + p_r = 1$.

References

1. G.A. Pagani, M. Aiello, The power grid as a complex network: a survey. Phys. A: Stat. Mech. Appl. **392**(11), 2688–2700 (2013)
2. Y. Koç, M. Warnier, P. Van Mieghem, R.E. Kooij, F.M. Brazier, A topological investigation of phase transitions of cascading failures in power grids. Phys. A: Stat. Mech. Appl. **415**, 273–284 (2014)
3. M. Rosas-Casals, S. Valverde, R.V. Solé, Topological vulnerability of the European power grid under errors and attacks. Int. J. Bifurc. Chaos **17**(07), 2465–2475 (2007)

4. P. Crucitti, V. Latora, M. Marchiori, A topological analysis of the italian electric power grid. Phys. A: Stat. Mech. Appl. **338**(1), 92–97 (2004)
5. S. Trajanovski, J. Martín-Hernández, W. Winterbach, P. Van Mieghem, Robustness envelopes of networks. J. Complex Netw. **1**(1), 44–62 (2013)
6. E. Bompard, E. Pons, D. Wu, Extended topological metrics for the analysis of power grid vulnerability. IEEE Syst. J. **6**(3), 481–487 (2012)
7. X. Wang, E. Pournaras, R.E. Kooij, P. Van Mieghem, Improving robustness of complex networks via the effective graph resistance. Eur. Phys. J. B **87**(9), 221 (2014)
8. P. Van Mieghem, K. Devriendt, H. Cetinay, Pseudoinverse of the laplacian and best spreader node in a network. Phys. Rev. E **96**(3), 032311 (2017)
9. R.D. Zimmerman, C.E. Murillo-Sánchez, R.J. Thomas, Matpower: steady-state operations, planning, and analysis tools for power systems research and education. IEEE Trans. Power Syst. **26**(1), 12–19 (2011)
10. H. Cetinay, F.A. Kuipers, P. Van Mieghem, A topological investigation of power flow. IEEE Syst. J. (2016)
11. D. Van Hertem, J. Verboomen, K. Purchala, R. Belmans, W. Kling, Usefulness of DC power flow for active power flow analysis with flow controlling devices, in *Proceedings IET ACDC'06* (2006), pp. 58–62
12. Y. Koç, M. Warnier, R.E. Kooij, F.M. Brazier, An entropy-based metric to quantify the robustness of power grids against cascading failures. Saf. Sci. **59**, 126–134 (2013)
13. H. Cetinay, S. Soltan, F.A. Kuipers, G. Zussman, P. Van Mieghem, Comparing the effects of failures in power grids under the ac and dc power flow models. IEEE Trans. Netw. Sci. Eng. (2017)
14. P. Van Mieghem, *Graph Spectra for Complex Networks* (Cambridge University Press, Cambridge, 2010)

Chapter 3
Grid Awareness Under Normal Conditions and Cyber-Threats

Matija Naglic, Arun Joseph, Kaikai Pan, Marjan Popov,
Mart van der Meijden and Peter Palensky

Abstract The situational grid awareness is becoming increasingly important for power system operations due to smaller operational margins, wide range of uncertainties entailed by renewables and highly critical infrastructure failures due to potential cyber-attacks. In this chapter, we look at some of the state-of-the-art technologies to monitor events in a power system under normal operating condition, followed by detection algorithms for regular business risk events, such as faults and equipment failures, and finally, we look into methods for quantifying vulnerability of under the rare and men-orchestrated cyber-attacks. First, we outline an architecture of a central piece of today's grid awareness system, Wide Area Monitoring, Protection and Control technology. Next, we review an event detection method used to identify and record faults and failures in the grid. Finally, we present a method for vulnerability assessment of grids under cyber-attacks.

M. Naglic (✉) · A. Joseph · K. Pan · M. Popov · M. van der Meijden · P. Palensky
Faculty of Electrical Engineering, Mathematics and Computer Science,
Delft University of Technology, Delft, The Netherlands
e-mail: m.naglic@tudelft.nl

A. Joseph
e-mail: arun.joseph@tudelft.nl

K. Pan
e-mail: k.pan@tudelft.nl

M. Popov
e-mail: m.popov@tudelft.nl

M. van der Meijden
e-mail: m.a.m.m.vandermeijden@tudelft.nl

P. Palensky
e-mail: p.palensky@tudelft.nl

© Springer Nature Switzerland AG 2019
P. Palensky et al. (eds.), *Intelligent Integrated Energy Systems*,
https://doi.org/10.1007/978-3-030-00057-8_3

3.1 Wide Area Monitoring, Protection and Control

The design, control and operation of Electric Power System (EPS) are undergoing significant changes. The environmental policies [1] are driving the society towards the zero-carbon emission by 2050 with energy production based only on renewable energy sources. The future EPS are expected to be highly interconnected (AC and DC grids) with significantly changed dynamics and dominated by intermittent renewable energy sources, typically integrated using power electronic devices. Additionally, electric vehicles as a sustainable alternative to internal combustion engine bring new patterns and challenges in energy supply. The new dynamics in energy demand and supply, unexpected disturbances, protection maloperation, and inadequate control schemes can cause system instabilities, leading to cascading failures and potentially catastrophic blackouts. A common reason for the occurrence of blackouts is lack of immediate coordinated protection and control when the EPS becomes unstable [2–4]. The conventional protection and control schemes are not designed to cope with the new local and wide area fast dynamics and states (conditions), imposed by future EPS. In the recent years, the Smart Grid technological advances in terms of sophisticated Intelligent Electronic Devices (IED), fast and reliable telecommunication links, and increased computational capacities have created new opportunities for design of next-generation monitoring, protection, and control schemes. As a result, the research and industry are driving towards advanced Wide Area Monitoring, Protection, and Control (WAMPAC) system [5]. The WAMPAC system is favorable to ensure efficient, more resilient, and secure operation of EPS by sophisticated utilisation of Smart Grid components in means of intelligent sensors, actuators, and state-of-the-art Information and Communications Technology (ICT). Typically, the WAMPAC system utilizes the advanced Synchronized Measurement Technology (SMT) [5–7] as a key building block to deliver time synchronized measurements (synchro-measurements) of electrical quantities from system-wide dispersed locations. Supported by precise time synchronization, the SMT comprises of intelligent electronic devices (IED) or Phasor Measurement Units (PMU), and Phasor Data Concentrators (PDC), connected over ICT infrastructure into a hierarchically organized network [8], as illustrated in Fig. 3.1.

3.1.1 WAMPAC-Ready Platform for Online Validation of Corrective Control Algorithms

Performance of a WAMPAC system and implemented applications is often mission critical and can in worst case scenario lead to system-wide blackout. Since actual SMT provided measurements can dramatically differ from conventional measurements, obtained from software simulations, it is prudent to extensively validate all emerging grid-saving applications under realistic conditions in an isolated and

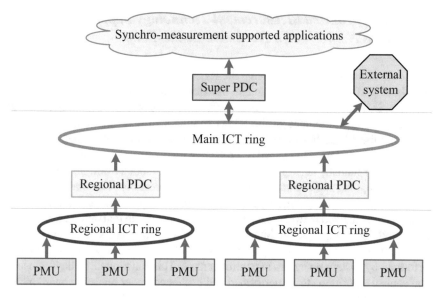

Fig. 3.1 Synchronized measurement technology infrastructure [8]

flexible simulation environment, before being implemented in an actual grid. Besides, communication delay, jitter, packet loss, available data throughput, and communication network reliability are the important factors that can significantly affect WAMPAC applications performance. To design a reliable and robust WAMPAC application, the abovementioned aspects need to be considered during WAMPAC application design and validation.

To serve this purpose, a simulation platform has been developed at the Delft University of Technology (TUD) [9], as a co simulation comprising of the SMT supported EPS model and the underlying ICT infrastructure. The presented WAMPAC-ready platform represents a cyber-physical simulator, due to the tight coupling between EPS and ICT, interacting with each other in a closed-loop manner, as illustrated in Fig. 3.2. The EPS component is represented by the RTDS real-time power system simulator with the integration of SMT components as HIL. In order to emulate the wide area ICT network, the open-source WANem network emulator and ns-3 network simulator are used as Software-in-the-Loop (SIL). To enable simplified design and online validation of WAMPAC applications, a MATLAB supported Synchro-measurement Application Development Framework (SADF) [8] has been in-house developed as SIL. In the following subsections we review the key parts of this simulation platform.

3.1.2 Synchronized Measurement Technology Supported EPS Network with Hardware-in-the-Loop

In order to simulate EPS phenomena, the RTDS real-time power system simulator, capable of performing electromagnetic transient simulations with a typical 50 μs time step, is used. The EPS network is built in a comprehensive RSCAD graphical user interface and simulated in the RTDS simulator in real-time. One of its main benefits is the possibility of HIL performance and compliance testing of control and measurement devices like PMUs, protective relays, circuit breakers, and control devices under realistic network conditions. Due to its modular design, it is suitable for fast prototyping and evaluation of diverse EPS applications. The simulated power system quantities (voltage and current waveforms, status signals and control commands) are exchanged between the RTDS and HIL devices in real time via numerous analog and digital input/output interfaces. In this way, feedback signals and control commands are used to change any set point or modify the EPS topology as typically performed by system operators in control centres. This functionality allows closed-loop control algorithms to be performed and evaluated under realistic conditions in a flexible simulation environment, before being implemented in the actual grids [8]. The presented platform in Fig. 3.2 consists of the SMT components needed to provide

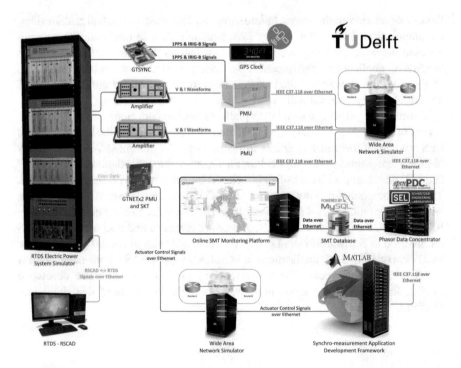

Fig. 3.2 WAMPAC-ready platform for online validation of corrective control algorithms [9]

Fig. 3.3 RTDS with SMT components

real time observability of grid dynamics based on the IEEE C37.118 [10, 11] standard. Two Alstom P847 IEDs are used, each capable of measuring two current and voltage channels, in total providing four synchrophasor streams. Hereby, two high-precision Omicron CMS-156 amplifiers are used to provide suitable signal levels for direct feeding to the actual PMUs (see Fig. 3.3). In addition, up to 24 embedded PMUs provided by the GTNET2x card can be used simultaneously, which adds up to 26 PMUs available in total. Moreover, to provide accurate time synchronization a GPS grand master clock is used to provide Inter-Range Instrumentation Group code B (IRIG-B) protocol based timestamp and 1 Pulse Per Second (1PPS) time signal to a GTSYNC card and the PMUs. Additionally, the clock provides IEEE 1588 Precision Time Protocol (PTP) time synchronization to Phasor Data Concentrators (PDCs) and SADF. OpenPDC and SEL software PDCs are used to aggregate and store separate online PMU data streams. The PDCs receive synchrophasor streams from all of the PMUs in the simulation, aligns synchrophasor measurements with the identical time-tags and aggregates them into a single synchrophasor data stream, forwarded to the MATLAB based Synchro-measurement Application Development Framework [8], as presented in Fig. 3.2. Additionally, synchrophasor measurements are simultaneously stored in the MySQL database, mainly for the online monitoring purpose and off-line analyses.

To simulate telecommunication characteristics, WANem network emulator is used, which in real-time emulates communication delay, jitter, packet loss, and available data throughput for each communication channel. In addition, ns-3 network simulator is used as SIL in order to investigate different ICT network related phenomena (link and node data-congestion, jitter, delay, packet drop etc.) and to assess the cyber-security vulnerabilities (false data injection, buffer overflows, device misconfiguration). By using the aforementioned ICT network simulators, each PMU stream and send control commands can be characterised with its own unique ICT

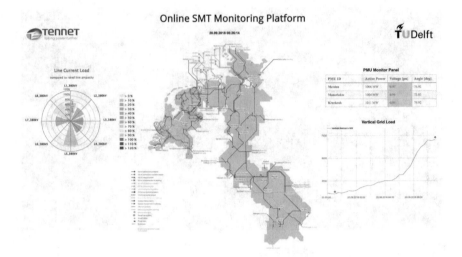

Fig. 3.4 Online SMT monitoring platform for real-time monitoring of EPS, viewed on tablet device [9]

performance characteristics that can imitate the actual ICT network conditions from the field. The presented platform can be utilized to analyse different cyber-attacks that could potentially endanger the safe operation of EPS. Different possible scenarios can be performed on the ICT components, data exchange, IEDs, and power hardware components.

3.1.3 Synchronized Measurement Technology Supported Monitoring Platform

As a part of the simulation platform [9], a web-based online SMT monitoring platform has been in-house developed (see Fig. 3.4) for the real-time monitoring of grid dynamics. The monitoring platform is mainly used for the online PMU measurements monitoring purposes, abnormal event detection and alarming, line load monitoring, offline analysis and data export. It connects to the online MySQL database, populated with the measurements from the OpenPDC. The web-platform is based on JavaScript and PHP programming languages and runs on top of the Linux operating system. Visualisation is based on HTML markup language with CCS and JQuery support. Due to the used protocols, it is cross-platform compatible (PC, tablet, smart phone) and intuitive to navigate (see Fig. 3.5).

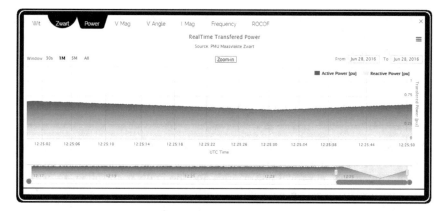

Fig. 3.5 Real-time monitoring of active and reactive power, measured by PMU device and viewed on tablet device [9]

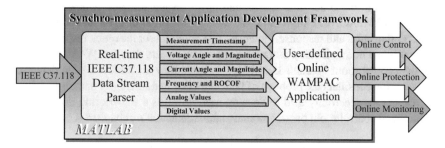

Fig. 3.6 MATLAB based Synchro-measurement Application Development Framework [8]

3.1.4 MATLAB Supported Synchro-measurement Application Development Framework

To enable a seamless integration between the SMT supported EPS and user-defined online applications, a MATLAB based Synchro-measurement Application Development Framework (SADF) has been developed (see Fig. 3.6) [8]. The SADF facilitates the real-time access and simplified use of the IEEE C37.118.2-2011 standard synchro-measurement data in the MATLAB programing environment. A MATLAB programing suite is a powerful software engineering toolbox supported the by numerous built-in mathematical, signal processing, and visualization functions. It is cross-platform compatible (Windows, Mac, Unix) and can be interfaced with C, C++, Java, Fortran, and Python programming languages. MATLAB has been a de facto programming language for research worldwide and has an extended online community support [12].

This concludes the presentation of the simulation platform. In the next section, we review a fast detection algorithm that can be used with such platform to detect and report faults to the system operator.

3.2   Fault Detection Using Synchro-measurements

The detection of a fault, its isolation and reconfiguration are important and challenging aspects power system operations and control. Components with complex dynamics interact in electrical power grids and the faults occur often. In extreme cases, they lead to cascading and highly undesirable events such as blackouts. Thus, quick fault detection is an imperative. The typical fault detection techniques are confined to the detection of a loss of a significant component (e.g., a transmission line, load, generation unit) using information from protection devices and SCADA (supervisory control and data acquisition) elements. The time scale of interest for detection is in the order of a few second up to several minutes. Modern systems, such as WAMPAC, offer high sampling rates (at the order of kHz) and potential for quicker detection and response. However, this potential for the quick response is not fully utilized due to lacking automation procedures, and thus far, WAMPAC relies on the control room operator knowledge to react after the fault has been detected. This section considers the problem of fault detection in a smart grid paradigm, where we expect the electricity grid to develop more towards a self-healing grid. Such a centralized control room based automated fault detection mechanism is proposed in this section.

The centralized control room should be equipped with tools to detect and react to undesired events, such as faults, in order to prevent cascading failures. The process starts with collecting synchro-measurements through the SMT system, as seen in Fig. 3.7. The measurements can be collected from the real system, or as in our case, from the EPS simulation, implemented in RTDS and interfaced through the GTNET cards. The second unit in Fig. 3.7 contains different fault detection algorithms that

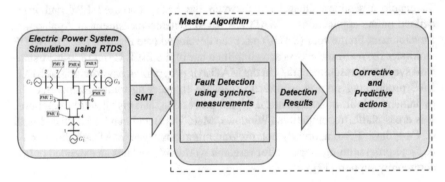

Fig. 3.7 Real-time platform

process the measurement data observing the faulty grid behavior. The third unit in the same figure consists of corrective and predictive measures. When working with synchro-measurements, if a fault is detected quickly, there often exists a sufficient time interval to predict the grid behavior before performing any action. To enable predictive actions, fast fault detection is of great importance. The further subsection illustrates the use of statistical methods as for fault detection.

3.2.1 Fault Detection Using Quickest Change Detection

In this method, the incremental changes in the net power injection at each bus are modeled as independent random variables, and the line outage is considered as a change event in the probability distribution of bus voltage phase-angle measurements obtained from SMT. As the real-time monitoring and operatability of the power systems is of great importance, extensive study has been made in line outage detection in the minimum possible time, particularly, Chen et. al in [13] proposed a method based on the theory of quickest change detection. In this subsection, a short description of the theory of QCD proposed in [13] and an application of this method for a 9 bus power system is depicted.

We consider a power system of N nodes denoted by a set $\chi = \{1, \ldots, N\}$, each of which correspond to a bus, and assume there are L edges. Let (m, n) denote the transmission line between buses m and n. Let $V_i(t)$ and $\theta_i(t)$ denote the voltage magnitude and phase angle respectively, at bus i. Also, let $P_i(t)$ and $Q_i(t)$ denote the net active and reactive power respectively, injected at bus i. In order to represent the quasi-steady-state behaviour of the power system, the real and reactive power balance components at each bus i can be written as:

$$P_i(t) = p_i \left(\theta_1(t), \ldots, \theta_N(t), V_1(t), \ldots, V_N(t) \right) \tag{3.1}$$

$$Q_i(t) = q_i \left(\theta_1(t), \ldots, \theta_N(t), V_1(t), \ldots, V_N(t) \right) \tag{3.2}$$

where $p_i(\cdot)$ and $q_i(\cdot)$ are functions used to represent the dependence on the network parameters. A linearized small signal power flow model is, however, considered and is used to perform the statistical test for change detection.

Considering the discretized version of the active and reactive power equations, we have

$$P_i[k] = p_i \left(\theta_1[k], \ldots, \theta_N[k], V_1[k], \ldots, V_N[k] \right) \tag{3.3}$$

$$Q_i[k] = q_i \left(\theta_1[k], \ldots, \theta_N[k], V_1[k], \ldots, V_N[k] \right) \tag{3.4}$$

where the time instant $t = k\Delta t$, $k = 1, 2, \ldots$ and $\Delta t > 0$. Defining $\Delta P_i[k] = P_i[k] - P_i[k-1]$, and $\Delta Q_i[k] = Q_i[k] - Q_i[k-1]$, and assuming that, for each bus i, $p_i(\cdot)$ and $q_i(\cdot)$ are continuously differentiable with respect to each θ_i and V_i at

$\theta_i[k]$ and $V_i[k]$, $\Delta P_i[k]$ and $\Delta Q_i[k]$ can be expressed using first order Taylor series expansion of (3.3) and (3.4) as

$$\Delta P_i[k] \approx \sum_{j=1}^{N} a_{ij}[2k]\Delta\theta_j[k] + \sum_{j=1}^{N} b_{ij}[2k]\Delta V_j[k] \tag{3.5}$$

$$\Delta Q_i[k] \approx \sum_{j=1}^{N} c_{ij}[2k]\Delta\theta_j[k] + \sum_{j=1}^{N} d_{ij}[2k]\Delta V_j[k] \tag{3.6}$$

where a_{ij} and b_{ij} are, respectively, the derivatives with respect to θ_j and V_j of p_i, and c_{ij} and d_{ij} are, respectively, the derivatives with respect to θ_j and V_j of q_i.

Based on the standard assumptions in the analysis of transmission systems, we have $a_{ij}[k] \gg b_{ij}[k]$, and $d_{ij}[k] \gg c_{ij}[k]$, and using the dc assumptions, $a_{ij}[k]$ becomes only the function of network alone, that is $a_{ij}[k] = a_{ij}$, and the analysis is presented only for $\Delta P_i[k]$, giving

$$\Delta P_i[k] \approx \sum_{j\in\chi, j\neq 1} a_{ij}\Delta\theta_j[k] \tag{3.7}$$

In matrix form, the above expression can be written as

$$\Delta P[k] \approx H_0 \Delta\theta[k] \tag{3.8}$$

where $\Delta P[k] \in \mathbb{R}^{N-1}$ and $\Delta\theta[k] \in \mathbb{R}^{N-1}$, the entries of which are $\Delta P_i[k]$ and $\Delta\theta_i[k]$ for $i \in \chi, i \neq 1$.

The physical fluctuations in the real power injection vector $\Delta P[k]$ is modeled as independent and identically distributed Gaussian random vectors of covariance Σ, that is $\Delta P[k] \sim \mathcal{N}(0, \Sigma)$. In terms of observation $\Delta\theta[k]$, we have

$$\Delta\theta[k] \approx M_0 \Delta P[k] \tag{3.9}$$

where $M_0 = H_0^{-1}$. Hence, under normal operation of the system, that is, prior to the line-outage event, $\Delta\theta[k] \sim f_0$, where

$$f_0 = \mathcal{N}(0, M_0\Sigma M_0^T). \tag{3.10}$$

Suppose a persistent outage occurs in line (m, n) at time $t = t_f$, where $(\gamma - 1)\Delta t \leq t_f < \gamma\Delta t$, for some $\gamma > 0$. In the event of outage, for $k \geq \gamma$, the matrix H_0 in (3.8) changes to $H_{(m,n)} = H_0 + \Delta H_{(m,n)}$, where $\Delta H_{(m,n)}$ is a perturbation matrix. Then the post-outage approximate power flow equation becomes

$$\Delta P[k] \approx H_{(m,n)}\Delta\theta[k] \tag{3.11}$$

and correspondingly

$$\Delta\theta[k] \approx M_{(m,n)}\Delta P[k] \tag{3.12}$$

where $M_{(m,n)} = H_{(m,n)}^{-1}$. Since H_0 has the same sparsity structure as the graph Laplacian of the network, the only non-zero terms in the matrix $\Delta H_{(m,n)}$ are $\Delta H_{(m,n)}[n,n] = -1/X_{(m,n)}$, $\Delta H_{(m,n)}[m,m] = -1/X_{(m,n)}$ and $\Delta H_{(m,n)}[m,n] = 1/X_{(m,n)}$. Thus, after the outage of the line (m,n), $\Delta\theta[k] \sim f_{(m,n)}^{\sigma}$, where

$$f_{(m,n)}^{\sigma} = \mathcal{N}(0, M_{(m,n)}\Sigma M_{(m,n)}^{T}). \tag{3.13}$$

for $k > \gamma$. It is assumed that the outaged line is not restored until a change is detected, that is, the change is persistent.

The goal of detecting line-outage has now became the problem of detecting the change in the probability distribution of the sequence $\{\Delta\theta[k]\}_{k\geq 1}$ as quickly as possible, while maintaining a certain level of detection accuracy, usually the probability of false alarm. Since it is assumed that the pre- and post-outage probability distribution functions f_0 and $f_{(m,n)}^{\sigma}$ are known, the popular Cumulative Sum (CuSum) algorithm can be used. In CuSum algorithm, a sequence of statistics is computed as

$$W_{(m,n)}[k+1] = \left(W_{(m,n)}[k] + \log\frac{f_{(m,n)}^{\sigma}(\Delta\theta[k])}{f_0(\Delta\theta[k])} \right)^{+} \tag{3.14}$$

where $W_{(m,n)}[0] = 0$ and the plus sign defined as $(x)^{+} = x$ if $x \geq 0$, otherwise $(x)^{+} = 0$. The CuSum algorithm hence declares a line outage in line (m,n) when the $W_{(m,n)}[k]$ computed for each line crosses a predetermined threshold A. The threshold A is adjusted according to the false alarm rate.

Denoting τ_C as the time at which the CuSum algorithm declares the line outage, we have

$$\tau_C = \inf\{k \geq 1 : W_{(m,n)}[k] > A\} \tag{3.15}$$

Since there are $L = 6$ buses, the line outage time is decided by choosing a time τ_{max}, at which the maximum of the L values of $W_{(m,n)}[k]$ crosses the threshold A. The relationship between A and the false alarm rate β is already established as $A = \log(L\beta)$ in [13]. Algorithms involving hypothesis testing rely on keeping the false alarm rate constant, which, in this case, is equivalent to keeping the quantity A constant. However, fixing the exact values of β is a matter of experience which falls in the hands of design engineers.

3.2.2 Other Statistical Algorithms for Change Detection

In the following section, two other statistical algorithms for change detection are described. They are explained as follows:

1. Meanshift test: Meanshift test is also known as one shot detection test wherein the distribution of the observations at a change point and before the change point is compared to detect the outage of line. Thus the algorithm detects a meanshift that occurs during the outage of a line ℓ between bus m and n using the test statistic:

$$W_\ell^\mu[k] = \log \frac{f_\ell^{CP}(\Delta\hat{\theta}[k])}{f_0(\Delta\hat{\theta}[k])} \qquad (3.16)$$

where $f_\ell^{CP}(\cdot)$ denote the distribution at the change point. The decision maker declares an outage when the test statistic of any of the L lines crosses their corresponding threshold A_ℓ. Thus the stopping time can be formally described as

$$\tau^\mu = \min_{\ell \in L} \left\{ \inf\{k \geq 1 : W_\ell^\mu[k] > A_\ell\} \right\}. \qquad (3.17)$$

2. Shewhart test: In QCD theory, Shewhart test is popularly used to perform change detection due to the easiness of implementation. A mild extension of Shewhart test for QCD in line outage system is proposed in [14], where the mean-shift and transient phenomenon is incorporated to the classical log-likelihood ratio between persistent post outage an pre-outage distributions. The statistic for outage of line ℓ is

$$W_\ell^{Sh}[k] = \max_{i \in \{0,1,\dots,T\}} \left\{ \log \frac{f_\ell^i(\Delta\hat{\theta}[k])}{f_0(\Delta\hat{\theta}[k])} \right\} \qquad (3.18)$$

where i indexes each of the T transient response periods. The stopping time for Shewhart's test is defined as

$$\tau^{sh} = \min_{\ell \in L} \left\{ \inf\{k \geq 1 : W_\ell^{sh}[k] > A_\ell\} \right\}. \qquad (3.19)$$

3.2.3 QCD Implementation in WECC Three-Machine Nine Bus System

The PMU placement for a WECC three-machine nine-bus system is shown in Fig. 3.8, with the PMUs placed in Buses 4, 5, 6, 7, 8 and 9. For the topology shown here, H_0 matrix is of order 6×6. By construction, H_0 matrix is a sparse matrix whose non-zero entries are calculated from reactance values of the line between buses. Figure 3.9 shows results of the case study. A 3-phase line to ground fault is simulated on the line between Bus 4 and Bus 5. The event starts at 0.4 s, with a fault duration chosen as 0.45 s and the Fig. 3.9 shows 1 s of system operation. The sequence of statistics $W_{(m,n)}$ from (3.14) is calculated using the real-time in-feed measurement data from SMT system. For the present study we have only considered the use of phase angle measurement values for calculation of sequence of the statistics. It can be noticed from Fig. 3.9 that with the chosen $A = 5000$ value the fault can be detected in a few milliseconds. This method was implemented as an algorithm and tested for different types of faults between the buses and scenarios of line outages, it was observed that by the proper selection of the threshold value A each fault can be differentiated and detected in very short time scales.

The algorithms, such as the one presented in this section, can be used to process synchro-measurements quickly, improving grid awareness to common equipment failures and faults, and enabling predictive response actions. In the next section, we look at the methodology for improving grid preparedness to cyber-attacks.

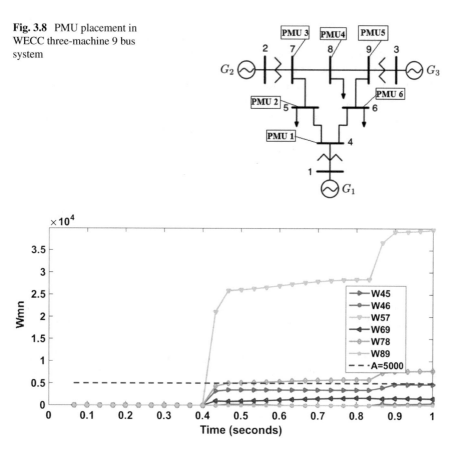

Fig. 3.8 PMU placement in WECC three-machine 9 bus system

Fig. 3.9 Sequence of statistics calculation for the outage of line 57

3.3   Vulnerability Assessment in Smart Grids Under Cyber-Attacks

The monitoring systems in power grids, e.g. SCADA and WAMPAC, depend heavily on proper operation of ICT infrastructure as the measurement data gets transported through the communication network to utility control center. However, cyber security vulnerabilities within the ICT infrastructures may allow attackers to manipulate SCADA and WAMPAC systems [15, 16]. Attackers can perform data integrity attacks by exploiting software vulnerabilities to inject false data on cyber components. System resources can also be rendered unavailable through denial of service (DoS) attacks by congesting the network. What makes things worse is that attackers can perform data integrity attacks and availability attacks in a coordinated way, i.e., combined data attacks. Combined data attacks expand the attack scenarios against power grids and even can bypass mitigation schemes for pure integrity or availability attacks [17].

Cyber-attacks on wide area monitoring systems would have great impact on power system reliability. Table 3.1, based on [18], lists typical data attacks and their impacts on several important applications like State Estimation (SE), Automatic Generation Control (AGC) and Special Protection Schemes (SPS). For instance, data attacks on SE could be achieved by compromising SCADA status and power flow measurements. Such attacks can result in a poor situation awareness of the power system and also lead to incorrect system operation leading in line overloads and market impacts in terms of uneconomical generation. Similarly, combined data attacks on AGC and SPS would impact the operational reliability and even cause cascading outages.

In order to increase the security of these systems, one needs analytic methods to first understand the vulnerabilities and then to validate or explore them with appropriate tools. However, analytic methods may have to ignore some details when modeling the heterogeneous intelligent power system, but could be used to create attack scenarios and guide the cyber security experiments on testbeds/tools. Thus, tools that integrate different cyber and physical components are needed to support the design and evaluation of cyber security issues of intelligent power system, from vulnerability analysis to attack impact assessment with empirical results. It should be noted that vulnerability and impact of attacks can be combined together in the notion of *risk*. Risk analysis methods and tools combining vulnerability and impact assessment for data attacks are needed to implement risk assessment methodologies [19]. In addition, the combination of analytic methods and numerical simulation could also contribute to develop mitigation schemes, e.g., a more robust algorithms/methods that combine system-theoretic and ICT-specific measures can be proposed to protect SCADA and WAMPAC against data attacks [20].

Table 3.1 Data attacks targeting electricity grid monitoring system (based on [18])

Attacks	Specific types	Access	Attack targets	Affected applications	Coordination	Possible impact
Integrity attack, availability attack, combined attack	Man-in-the-middle attack, buffer overflow attack, DDoS flooding attack	Via SCADA network, RTU	SCADA status and power flow measurements	State estimation	Space, same time	Line overloads, poor situation awareness, economic loss
		Via SCADA network, RTU	Frequency, tie-line flow measurements	Automatic generation control	Space, same time	Frequency imbalance, operational reliability
		Via WAMPAC network, PMU	PMU measurements	Special protection schemes	Space, staggered time	Operational reliability, blackout or even cascading outages

3.3.1   Analytic Vulnerability Assessment

3.3.1.1   Data Attacks and Vulnerability Assessment Problem

Vulnerability of intelligent power system to data attacks is usually quantified by computing attack resources needed by the attacker to corrupt the system and keep undetectability. The vulnerability assessment is presented through the notion of *security index*. This metric can quantify the attack resources and consequently the vulnerabilities of the system to attacks. The intelligent power grid is more vulnerable to attacks with small security metric since such attacks need less resources to be executed.

We take data attacks on SE as an example. The data collected from substations includes line power flow and bus power injection measurements. These m measurements are denoted by $z = [z_1, \ldots, z_m]^T$. We assume that a power system has $n + 1$ buses, and that there are n phase angles to be estimated excluding the reference angle under the DC power flow model, i.e., the system state $z = [x_1, \ldots, x_n]^T$. We can write $z = Hx + e$, where $H \in \mathbb{R}^{m \times n}$ is the constant Jacobian matrix and $e \in \mathbb{R}^m$ is the measurement noise vector. With the goal of perturbing the SE and further corrupting the applications in EMS, the attacker would gain access to the measurement data. The measurements under different attack scenarios from the view of SE can be presented as follows:

- Data integrity attack - also known as false data injection (FDI) attack, is able to change measurement values from z to $z + a$ where a is the *FDI attack vector*.

- Data availability attack - includes DoS or jamming attack which would make specific measurements unavailable to SE, i.e., $z_0 = (I - \text{diag}(d))z$ where $d \in \{0, 1\}^m$ is the *availability attack vector* and I is an identity matrix.
- Combined attack - combines the FDI and availability attack that makes the measurements from z to $(I - \text{diag}(d))z + a$ corrupted by a and d.

To formulate the security index, optimization problems with the objective specified as security index and constraints corresponding to the undetectability or impact conditions are proposed to characterize attacks with different attack vectors. Looking at the attack scenarios above, if the attacker corrupts certain measurements using FDI attack vector $a = Hc$, it can remain hidden from being detected by some built-in bad data detection schemes but perturb the current state to a degree of c. It's also shown in our recent work [17] that combined attacks can achieve the same target with the attack vector $a = (I - \text{diag}(d))Hc$. An illustrative security index can be introduced as the minimum number of measurements that need to be corrupted by the attacker, with the objective $\alpha_j := \min_{a,d} \ \|a\|_0 + \|d\|_0$ and constraints corresponding to the undetectability conditions for attack vectors.

3.3.1.2 Analytic Vulnerability Assessment Incorporating Communication Network Properties

The vulnerability assessment problem above shows the measurements/sensors that need to be manipulated by the attacker. This security metric suits for the cases that attacks arise from the level of sensors. A more interesting scenario is to look into the attacks from the level of communication networks since usually the attacker would explore vulnerabilities in the networks, e.g., compromising remote access points, obtaining access to corporate networks. The vulnerability assessment should consider the communication network. However, modeling the communication network in an analytic framework is challenging due to its complexity and heterogeneity. Here, the communication network properties of interest for security analysis are as follows:

- Communication topology;
- Routing schemes - the routing paths of packets / data;
- Communication latency - how the packets / data would be delayed in each communication infrastructure;
- Packet loss / missing data - the possibility of packet drop in each communication infrastructure.

In [17, 21] we introduced a method to deal with the first two properties that can be employed in the analytic vulnerability assessment. We introduced a binary vector called *routing vector* which denotes the communication nodes/links that each measurement traverses. Using the graph of the communication network and routing schemes for all the measurements, we can build a *routing matrix* stacked by the routing vectors. The routing matrix contains the information of communication topology and routing schemes. The network-aware security index can be introduced as the minimum number of communication nodes/links that need to be attacked and the constraints use the *routing vectors* to map the data attacks on measurements to attacks on the communication network. Thus the security index can illustrate the vulnerability of SE to data attacks on the communication network.

It should be noted that some ICT-specific security measures can be modeled in such security index problem. For instance, multi-path routing schemes can be described using *routing vectors*. Data authentication can be implemented by adding constraints to indicate measurements that originate from the node protected by authentication. More cyber properties of communication networks, latency and packet loss, can also be considered as the factors that impact the construction of the security index problem.

These security indexes from the analytic vulnerability assessment specify the measurements/communication infrastructures to be attacked, thus could be used to create attack scenarios for simulation. However, such security index do not consider the attack impact on the physical system operation. In fact, data attacks with the same security metrics could have considerable different impact. Co-simulation could offer the capabilities to look into the attack impact and provide empirical results to validate and contribute in developing mitigation measures.

3.3.2 Co-simulation Supporting Vulnerability and Impact Analysis

3.3.2.1 Co-simulation Framework for Risk Assessment

In our papers [22, 23], we discussed co-simulation of intelligent power grids and its potential applications for cyber security analysis. A co-simulation framework is an integrated environment including simulators of power system, communication network and EMS applications. Under the co-simulation, the communication model can be implemented as a hierarchical one that is close to reality. From co-simulation, the attack scenarios from the analytic vulnerability assessment can be validated and the attack impact can be explored from simulation results. The risk of system to data attacks can be assessed incorporating both vulnerability and attack impact.

The co-simulation framework is shown in Fig. 3.10 and is implemented on top of the integration of power system, communication network and control/application simulators. This platform should: (1) be modular, extensible and flexible to simulate communication networks; (2) allow implementation of attacks and mitigation schemes.

3.3.2.2   Co-simulation Setup: Tools, Integration and Attacks Modeling

Here we present our co-simulation platform as shown in Fig. 3.11 and further details are referred to [21]. This co-simulation platform is implemented with three tools: DIgSILENT PowerFactory for the power system, OMNeT++ for the communication network, and Matlab/Matpower for the EMS algorithms. OMNeT++ is selected because it is a generic simulation engine, open source and it allows plug-n-play through NED editor and integration to external devices. The *scheduler* of OMNeT++ is customized as the master algorithm and external interface for integration with PowerFactory and Matlab/Matpower (Fig. 3.11).

1. Power system simulator: DIgSILENT PowerFactory is used to conduct a quasi-static power flow simulation. PowerFactory's Python API is used to create a script that controls the execution of the simulation. The same script implements the interface with OMNeT++. Real time execution is achieved by synchronizing the power flow calculations with the system clock. The script sends measurements to OMNeT++ every time period (set to be 5 s), but it can expect generator set points at any time.
2. Communication network simulator: OMNeT++ is used for discrete-event based communication network simulation. A custom OMNeT++ *scheduler* is built to enable data exchange with PowerFactory and Matlab over TCP/IP sockets and run the OMNeT++ in real-time. The communication nodes such as RTUs, modems and routers are built using the modules in OMNeT++ represent the LAN (local area network) of a substation. Besides, there are two kinds of communication links/channels: (1) channel of the LAN between RTUs and modems; (2) channel of the WAN (wide area network) between routers.

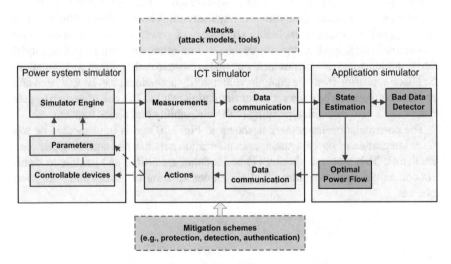

Fig. 3.10 Co-simulation based cyber security risk assessment framework

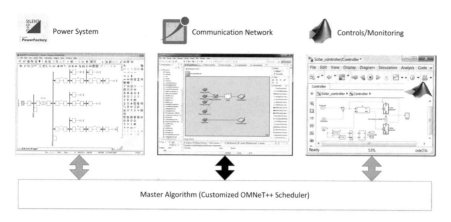

Fig. 3.11 A co-simulation platform setup

3. EMS algorithm: Matpower has been used to simulate the EMS applications in Matlab, including state estimation (with bad data detection) and optimal power flow algorithms. A script is implemented to exchange data with OMNeT++ scheduler over TCP/IP sockets and store measurements into a data pool. The State Estimation module uses the latest measurements from data pool to create a snapshot of estimated power flow. The Optimal Power Flow module uses load estimates from State Estimation to perform optimal power flow calculation and sends commands of generator set points to PowerFactory through OMNeT++.

The integration of simulators is implemented as follows: data is exchanged between PowerFactory, OMNeT++ and Matlab via TCP/IP sockets using the ASN.1 protocol. On the PowerFactory side, data exchange is implemented in the Python script that controls the simulator execution, while on the OMNeT++ side, data exchange is implemented through a custom scheduler. This scheduler acts as the "master" to coordinate the co-simulation, handle the data exchanges with PowerFactory and Matlab, and also run the OMNeT++ in a real-time mode.

As discussed in the previous section, an attacker can manipulate the measurements by injecting false data, making it unavailable or both. After accessing a router, the attacker can launch data integrity and availability attacks on all the data traveling through it by executing a *man-in-the-middle attack*. By jamming, DoS or physical attack, the attacker can block measurements in communication links. The attack scenarios from the analytic vulnerability assessment can be conducted. The results from network-aware security index problem is used to choose the routers to be attacked. This attack is implemented in OMNeT++ by changing the behavior of the router in case it is accessed by the attacker.

It should be noted that these types of attacks can be modeled based on some attack "libraries". For instance, the NETA framework [24] can be used and further developed to add attack modules in the simulation model. Moreover, attacks like Man-in-the-middle attack, DDoS flooding attack, can be implemented by using available tools (e.g., Ettercap suite, Tribe Flood Network tool) and integrating them into the co-simulation. Besides, the mitigation schemes for attacks, e.g., protection and authentication, can also be implemented by adding the configurations to the modules in OMNeT++.

References

1. Long Term Global Goals for 2050, Climate Action Network International, http://www.climatenetwork.org/sites/default/files/can_position-long_term_global_goals_for_2050.pdf. Accessed 4 June 2018
2. Final report of the investigation committee on the 28 September 2003 blackout in Italy, TUCE (2004)
3. U.S.-Canada Power System Outage Task Force, in *Final Report on the August 14, 2003 Blackout in the United States and Canada: Causes and Recommendations* (Natural Resources Canada, 2004)
4. International Conference on Large High Voltage Electric Systems and Conférence Internationale des Grands Réseaux Electriques à Haute Tension. *Wide Area Monitoring and Control for Transmission Capability Enhancement* (Cigré, 2007)
5. V. Terzija, G. Valverde, D. Cai, P. Regulski, V. Madani, J. Fitch, S. Skok, M.M. Begovic, A. Phadke, Wide-area monitoring, protection, and control of future electric power networks. Proc. IEEE **99**(1), 80–93 (2011)
6. F. Aminifar, M. Fotuhi-Firuzabad, A. Safdarian, A. Davoudi, M. Shahidehpour, Synchrophasor measurement technology in power systems: Panorama and state-of-the-art. IEEE Access **2**, 1607–1628 (2014)
7. G.A. Phadke, J.S. Thorp, Synchronized Phasor Measurements and Their Applications (Springer International Publishing, Berlin, 2017)
8. M. Naglic, M. Popov, M.A.M.M. Meijden, V. Terzija, Synchro-measurement application development framework: an IEEE standard C37.118.2-2011 supported MATLAB library. IEEE Trans. Instrum. Meas. 1–11 (2018)
9. M. Naglic, I. Tyuryukanov, M. Popov, M. van der Meijden, V. Terzija, WAMPAC-ready platform for online evaluation of corrective control algorithms, in *Mediterranean Conference on Power Generation, Transmission, Distribution and Energy Conversion (MedPower 2016)* (Institution of Engineering and Technology, 2016)
10. IEEE, *Std C37.118.1-2011 (Revision of IEEE Std. C37.118-2005): IEEE Standard for Synchrophasor Measurements for Power Systems* (IEEE, 2011)
11. IEEE, *Std. C37.118.1a, IEEE Standard for Synchrophasor Measurements for Power Systems - Amendment 1: Modification of Selected Performance Requirements* (IEEE, 2014)
12. Matlab Central
13. Y.C. Chen, T. Banerjee, A.D. Domínguez-García, V.V. Veeravalli, Quickest line outage detection and identification. IEEE Trans. Power Syst. **31**(1), 749–758 (2016)
14. G. Rovatsos, X. Jiang, A.D. Domínguez-García, V.V. Veeravalli, Comparison of statistical algorithms for power system line outage detection in 2016, in *IEEE International Conference on Acoustics, Speech and Signal Processing (ICASSP)* (2016), pp. 2946–2950
15. T.M. Thomas, S. Abu-Nimeh, Lessons from stuxnet. Computer **44**(4), 91–93 (2011)
16. A. Hahn, A. Ashok, S. Sridhar, M. Govindarasu, Cyber-physical security testbeds: architecture, application, and evaluation for smart grid. IEEE Trans. Smart Grid **4**(2), 847–855 (2013)
17. K. Pan, A.M.H. Teixeira, M. Cvetkovic, P. Palensky, Combined data integrity and availability attacks on state estimation in cyber-physical power grids, in *Proceedings of the IEEE International Conference on Smart Grid, Communications (SmartGridComm)* (2016), pp. 271–277
18. A. Ashok, A. Hahn, M. Govindarasu, Cyber-physical security of wide-area monitoring, protection and control in a smart grid environment. J. Adv. Res. **5**(4), 481–489 (2014)

19. K. Pan, A. Teixeira, M. Cvetkovic, P. Palensky, Cyber risk analysis of combined data attacks against power system state estimation. IEEE Trans. Smart Grid. https://doi.org/10.1109/TSG.2018.2817387
20. M. Findrik, P. Smith, J.H. Kazmi, M. Faschang, F. Kupzog, Towards secure and resilient networked power distribution grids: process and tool adoption, in *2016 IEEE International Conference on Smart Grid Communications (SmartGridComm)*, pp. 435–440, 2016
21. K. Pan, A. Teixeira, C.D. López, P. Palensky, Co-simulation for cyber security analysis: data attacks against energy management system, in *Accepted in 8th IEEE International Conference on Smart Grid Communications (SmartGridComm)* (2017)
22. P. Palensky, A.A. Van Der Meer, C.D. López, A. Joseph, K. Pan, Cosimulation of intelligent power systems: fundamentals, software architecture, numerics, and coupling. IEEE Indust. Electron. Mag. **11**(1), 34–50 (2017)
23. P. Palensky, A. van der Meer, C. Lopez, A. Joseph, K. Pan, Applied cosimulation of intelligent power systems: implementing hybrid simulators for complex power systems. IEEE Indust. Electron. Mag. **11**(2), 6–21 (2017)
24. L. Sánchez-Casado, R.A. Rodríguez-Gómez, R. Magán-Carrión, G. Maciá-Fernández, Neta: evaluating the effects of network attacks. Manets as a case study, in *Advances in Security of Information and Communication Networks* (Springer, 2013), pp. 1–10

Part III
Simulations of Future Power Systems

Chapter 4
Globalized Newton–Krylov–Schwarz AC Load Flow Methods for Future Power Systems

Domenico Lahaye and Kees Vuik

Abstract The load flow equations express the balance of power in an electrical power system. The power generated must equal the power consumed. In the AC time-harmonic case, the load flow equations are non-linear in the voltage phasors associated with the nodes in the network. The development of future power systems urgently requires new, highly efficient and robust load flow solvers. In this contribution we aim at making the following three scientific contributions. We first show that the use of a globalization procedure is required to ensure the convergence of a Newton load flow simulation of a stressed network. Such operational conditions are more likely to occur in the future. We subsequently show that the use of an inexact Newton–Krylov method results in faster computations. We employ Quotient Minimal Degree (QMD) as a matrix reordering method, incomplete LU factorization (ILU) as a preconditioner, Generalized Minimal Residual (GMRES) as a Krylov acceleration, and the Dembo-Steihaus strategy to defined the accuracy of the linear solver at each Newton iteration. We finally show the results of iterative solution algorithms that allow to exploit the decomposition of a network into subnetworks. Decompositions with and without overlapping nodes are tested.

4.1 Introduction

The development of new power system architectures motivates the parallel development of new computational tools. These new tools will have to be able to give fast and reliable answers in challenging and unforeseen conditions. Emerging electrical power networks will be operated closer to limits specified in the design, allow bi-directional power flow, accommodate renewable sources and electrical vehicles and merge with gas and heat networks. More simulations will be required to decide on the type of control action to take in the grid. The size of the problems is expected

D. Lahaye (✉) · K. Vuik
Faculty of Electrical Engineering, Mathematics and Computer Science,
Delft University of Technology, Delft, The Netherlands
e-mail: d.j.p.lahaye@tudelft.nl

K. Vuik
e-mail: c.vuik@tudelft.nl

© Springer Nature Switzerland AG 2019   79
P. Palensky et al. (eds.), *Intelligent Integrated Energy Systems*,
https://doi.org/10.1007/978-3-030-00057-8_4

to grow with the extend of geographical interconnections and the level of detail in the network description. Computationally efficient and robust methods will therefore have to be developed.

The load-flow equations describe the flow of electrical power in an electrical power system [1–6]. In a time-harmonic formulation, these equation on non-linear in the complex-valued nodal voltage phasors. Various commercial and public domain software packages that implements solvers for these equations is given on the URL [7]. The Jacobian matrix of partial derivatives can be computed analytically. Newton's method [8] is therefore the solution method of choice. Newton's method is however only guaranteed to converge for initial guesses that are sufficiently close to the solution. In traditional power system analysis, the load-flow solution has an amplitude close to one and an angle close to zero degrees (in per unit values). By specifying an initial guess with unit amplitude and zero angle (referred to as a flat start), Newton's method is likely to converge. In the future power systems, more frequent and large deviations from the nominal solution of the load flow equations are expected to occur. This will lead to nodal voltage and line overloads. In these cases, Newton's method is no longer likely to converge with flat start initial guesses. We show that in these case the convergence can be guided by globalizing the Newton method. In a globalized form, the Newton's method is augmented with auxiliary procedures that ensure convergence from initial guesses that lie further way from the solution.

The Newton method required solving a linear system with a large and sparse Jacobian system at each iteration. Preconditioned Krylov subspace methods are ideally suited to solve these systems [9]. We give examples of how load-flow equations can efficiently be solved using preconditioned Krylov subspace methods. We show numerical results for the load flow solver implemented in the PETSc library [10]. The test cases are taken from the examples provided with the MatPower package [11]. We in particular show how the additive Schwarz method that decomposes the network with and without overlapping nodes can be employed.

This chapter is structured as follows. In Sect. 4.2 the concepts of voltage, current and power for the steady state description of a power system are introduced. In Sect. 4.3 the power system is represented as a graph with generator and load nodes and interconnecting edges. In Sect. 4.4 the load flow equations and the method of Newton to solve them are described. In Sect. 4.5 an example of the computational efficiency of the Newton–Krylov method is given. In Sect. 4.6 an example of solving the load flow equations using the Schwarz method is given. Concluding remarks are made in Sect. 4.7.

4.2 Power Systems

A system of hardware that provides for the generation of power, and the transmission to the consumers, is called a Power System. This section contains an introduction to voltage, current, and power, for the steady state of a power system operating on alternating current (AC).

4.2.1 Voltage and Current

The voltage and current, in a power system in steady state, can be assumed to be
sinusoidal functions of time with constant frequency ω. It is conventional to use the
cosine function to describe these quantities, i.e.,

$$v(t) = V_{\max} \cos(\omega t + \delta_V) = \Re\left(V_{\max} e^{j\delta_V} e^{j\omega t}\right),$$
$$i(t) = I_{\max} \cos(\omega t + \delta_I) = \Re\left(I_{\max} e^{j\delta_I} e^{j\omega t}\right),$$

where j is the imaginary unit, and $\Re$ is the operator that takes the real part.

The complex numbers $V = V_{\max} e^{j\delta_V}$ and $I = I_{\max} e^{j\delta_I}$ are called the phasor representation of the voltage and current respectively, and are used to represent the
voltage and current in circuit theory. In power system theory, instead the effective
phasor representation is used:

$$V = |V| e^{j\delta_V}, \quad \text{with } |V| = \frac{V_{\max}}{\sqrt{2}}, \tag{4.1}$$

$$I = |I| e^{j\delta_I}, \quad \text{with } |I| = \frac{I_{\max}}{\sqrt{2}}. \tag{4.2}$$

Note that $|V|$ and $|I|$ are the RMS values of $v(t)$ and $i(t)$, and that the effective
phasors differ from the circuit theory phasors by a factor $\sqrt{2}$.

As we are dealing with power system theory, we use V and I to denote the effective
voltage and current phasors, as defined above. Further, as usual in the treatment of
power systems, we use per unit quantities, and represent the balanced three-phase
network of a power system as its equivalent single-phase system. For more details
see for example [4] or [3].

4.2.2 Complex Power

Assume that t is chosen such that the voltage is $v(t) = V_{\max} \cos(\omega t)$, and the current
is $i(t) = I_{\max} \cos(\omega t - \phi)$. The value $\phi = \delta_V - \delta_I$ is called the power factor angle,
and $\cos\phi$ the power factor. Then the instantaneous power $p(t)$ is given by

$$
\begin{aligned}
p(t) &= v(t)\, i(t) \\
&= \sqrt{2}\,|V| \cos(\omega t)\, \sqrt{2}\,|I| \cos(\omega t - \phi) \\
&= 2\,|V|\,|I| \cos(\omega t) \cos(\omega t - \phi) \\
&= 2\,|V|\,|I| \cos(\omega t) \cos(\omega t - \phi) \\
&= 2\,|V|\,|I| \cos(\omega t) \left[\cos\phi \cos(\omega t) + \sin\phi \sin(\omega t)\right] \\
&= |V|\,|I| \left[2 \cos\phi \cos^2(\omega t) + 2 \sin\phi \sin(\omega t) \cos(\omega t)\right]
\end{aligned}
$$

$$\begin{aligned}
&= |V|\,|I|\cos\phi\left[2\cos^2(\omega t)\right] + |V|\,|I|\sin\phi\left[2\sin(\omega t)\cos(\omega t)\right]\\
&= |V|\,|I|\cos\phi\left[1 + \cos(2\omega t)\right] + |V|\,|I|\sin\phi\left[\sin(2\omega t)\right]\\
&= P\left[1 + \cos(2\omega t)\right] + Q\left[\sin(2\omega t)\right],
\end{aligned}$$

where $P = |V|\,|I|\cos\phi$, and $Q = |V|\,|I|\sin\phi$.

Thus the instantaneous power is the sum of a unidirectional component, that is sinusoidal with average value P, and amplitude P, and a component of alternating direction, that is sinusoidal with average value 0, and amplitude Q. Note that integrating the instantaneous power over one time period $T = \frac{2\pi}{\omega}$ gives

$$\int_0^T p(t)\,dt = P.$$

The magnitude P is called the active power, or also real power or average power, and is measured in W (watts). The magnitude Q is called the reactive power, or also imaginary power, and is measured in VAr (voltamperes reactive).

Using the complex representation of voltage and current, we can write

$$P = |V|\,|I|\cos\phi = \Re\left(|V|\,|I|\,e^{j(\delta_V - \delta_I)}\right) = \Re\left(V\overline{I}\right),$$
$$Q = |V|\,|I|\sin\phi = \Im\left(|V|\,|I|\,e^{j(\delta_V - \delta_I)}\right) = \Im\left(V\overline{I}\right),$$

where $\overline{I}$ is the complex conjugate of I. Thus we can extend Joule's law to AC circuits, as

$$S = V\overline{I}. \tag{4.3}$$

Note that strictly speaking VA and VAr are the same unit as W, however it is very useful to use descriptive units to distinguish between the measured quantities.

4.2.3 Impedance and Admittance

Impedance is the extension of the notion resistance, to AC circuits, and is thus a measure of opposition to a sinusoidal current. The impedance is denoted by $Z = R + jX$, and measured in ohms (Ω). We call $R \geq 0$ the resistance, and X the reactance. If $X > 0$ the reactance is called inductive and we can write $jX = j\omega L$, where $L > 0$ is called the inductance. If $X < 0$ the reactance is called capacitive and we write $jX = \frac{1}{j\omega C}$, where $C > 0$ is called the capacitance.

The admittance $Y = G + jB$ is the inverse of impedance, i.e., $Y = \frac{1}{Z} = \frac{R}{|Z|^2} - j\frac{X}{|Z|^2}$, and is measured in siemens (S). We call $G = \frac{R}{|Z|^2} \geq 0$ the conductance, and $B = -j\frac{X}{|Z|^2}$ the susceptance.

The voltage drop over an impedance Z is equal to $V = ZI$. This is the extension of Ohm's law to AC circuits. Alternatively, using the admittance, we can write

$$I = YV. \tag{4.4}$$

The power consumed by the impedance is $S = V\overline{I} = ZI\overline{I} = |I|^2 Z = |I|^2 R + j |I|^2 X$.

4.2.4 Kirchhoff's Circuit Laws

To calculate the voltage and current in an electrical circuit, we use Kirchoff's circuit laws.

Kirchhoff's current law (KCL)
At any point in the circuit that does not represent a capacitor plate, the sum of currents flowing towards that point is equal to the sum of currents flowing away from that point, i.e., $\sum_k I_k = 0$.

Kirchhoff's voltage law (KVL)
The directed sum of the electrical potential differences around any closed circuit is zero, i.e., $\sum_k V_k = 0$.

4.3 Power System Model

Power systems are modeled as a network of buses (nodes) and lines (edges). At each bus i four electrical magnitudes are of importance:

$|V_i|$, the voltage amplitude,
δ_i, the voltage phase angle,
P_i, the injected active power,
Q_i, the injected reactive power.

Each bus can consist of a number of electrical devices. The bus is named according to the electrical magnitudes specified by its devices,

generator bus or PV-bus: P_i and $|V_i|$ specified, Q_i and δ_i unknown,
load bus or PQ-bus: P_i and Q_i specified, $|V_i|$ and δ_i unknown.

4.3.1 Generators, Loads, and Transmission Lines

Generally, a physical generator has P and $|V|$ controls and thus specifies these magnitudes. Likewise, a load will have specified negative injected active power P, and specified reactive power Q. However, the name of the bus does not necessarily

indicate what type of devices it consists of. A wind turbine, for example, is a generator
but does not have PV controls. A wind turbine is instead modeled as a load, with
positive injected active power P. If a PV generator and a PQ load are connected to
the same bus, the result is a PV-bus with a voltage amplitude equal to that of the
generator, and an active power equal to the sum of the active power of the generator
and the load. Also there may be buses without a generator or load connected, such
as transmission substations, which are modeled as load with $P = Q = 0$.

In any practical power system there are system losses. These losses have to be
taken into account, but since they depend on the power flow they are not known in
advance. Therefore one generator bus has to be assigned to supply these unknown
losses. This bus is generally called the slack bus. Obviously it is no longer possible
to specify the real power P for this bus. Instead the voltage magnitude $|V|$ and angle
δ are specified. Note that δ does not have any practical meaning, it is merely the
reference phase to which the other phase angles are related. As such, for the slack
bus it is generally specified that $\delta = 0$.

Lines are the network representation of the transmission lines, that connect the
buses in the power system. From a modeling viewpoint, lines define how to relate
buses through Kirchhoff's circuit laws. Lines can incur losses on the transported
power and must be modeled as such.

A transmission line from bus i to bus j has some impedance. The total impedance
over the line is modeled as a single impedance z_{ij} of the line. The admittance of
that line is $y_{ij} = \frac{1}{z_{ij}}$. Further there is a shunt admittance from the line to the neutral
ground. The total shunt admittance y_s of the line is modeled to be evenly divided
between bus i and bus j.

It is usually assumed that there is no conductance from the line to the ground.
This means that the shunt admittance is due only to the electrical field between line
and ground, and is thus a capacitive susceptance, i.e., $y_s = jb_s$, with $b_s > 0$. For this
reason, the shunt admittance y_s is also sometimes referred to as the shunt susceptance
b_s. See also the notes about modeling a shunt in Sect. 4.3.2. For details on how to
calculate y_s for a given line, we refer to [4].

4.3.2 Shunts, Tap Transformers, and Phase Shifters

Three other devices, that are also commonly found in power systems, are shunts,
tap transformers and phase shifters. Shunt capacitors can be used to inject reactive
power, resulting in a higher node voltage, whereas shunt inductors consume reactive
power, thus lowering the node voltage.

A shunt is modeled as a reactance $z_s = jx_s$ between the bus and the ground. The
shunt admittance is thus $y_s = \frac{1}{z_s} = -j\frac{1}{x_s} = jb_s$. If $x_s > 0$ the shunt is inductive, and
if $x_s < 0$ the shunt is capacitive. Note that the shunt susceptance b_s has the opposite
sign of the shunt reactance x_s.

A tap transformer is a transformer that can be set to different turns ratios. The taps are the connection points along the transformer winding on one side, that allow a certain number of turns to be selected. Tap transformers are generally used to control the voltage magnitude, dealing with fluctuating industrial and domestic demands, or with the effects of switching out a circuit for maintenance. Note that transformers without taps do not need to be modeled specifically, as the per unit system reduces their numerical effect to that of a series impedance.

Phase shifters are devices that can change the voltage phase angle, while keeping the voltage magnitude constant. As such they can be used to control the active power.

The transformer ratio is given by $T : 1$. For a tap transformer T is a positive real number, typically between 0.8 and 1.2. In the case of a phase shifter, T is a rotation in the complex plane, i.e., $T = e^{j\delta_T}$, where δ_T is the phase shift.

4.3.3 Admittance Matrix

The admittance matrix Y, is a matrix that relates the injected current I at each bus to bus voltages V, such that

$$I = YV, \tag{4.5}$$

where I is the vector of injected currents at each bus, and V is the vector of bus voltages. This is in fact Ohm's law (4.4) in matrix form. As such we can also define the impedance matrix $Z = Y^{-1}$.

To calculate the admittance matrix Y, we look at the injected current I_i at each bus i. If $I_i > 0$ power is generated, if $I_i < 0$ power is consumed, and if $I_i = 0$ there is no current injected, for example at a transmission substation. Let I_{ij} denote the current flowing from bus i, in the direction of bus j, or to the ground in case of a shunt. Applying Kirchoff's current law now gives

$$I_i = \sum_k I_{ik}. \tag{4.6}$$

Let y_{ij} denote the admittance of the line between bus i and j, with $y_{ij} = 0$ if there is no line between these buses. For a simple transmission line from bus i to bus j, without shunt admittance, Ohm's law states that

$$I_{ij} = y_{ij} \left(V_i - V_j \right), \text{ and } I_{ji} = -I_{ij}, \tag{4.7}$$

or in matrix notation:

$$\begin{bmatrix} I_{ij} \\ I_{ji} \end{bmatrix} = y_{ij} \begin{bmatrix} 1 & -1 \\ -1 & 1 \end{bmatrix} \begin{bmatrix} V_i \\ V_j \end{bmatrix}. \tag{4.8}$$

Note that, if a power system would consist of such simplified lines only, the admittance matrix for that system is a Laplacian matrix given by

$$Y_{ij} = \begin{cases} \sum\limits_{k \neq i} y_{ik} & \text{if } i = j, \\ -y_{ij} & \text{if } i \neq j. \end{cases} \tag{4.9}$$

Indeed, then we have

$$I_i = \sum_k I_{ik} = \sum_k y_{ik} (V_i - V_k) = \sum_{k \neq i} y_{ik} V_i - \sum_{k \neq i} y_{ik} V_k = \sum_k Y_{ik} V_k = (YV)_i .$$
$$\tag{4.10}$$

Now suppose there is a shunt s connected to bus i. Then, according to Eq. (4.6), an extra term I_{is} is added to the injected power I_i. It is clear that

$$I_{is} = y_s (V_i - 0) = y_s V_i. \tag{4.11}$$

This means that an extra term y_s has to be added to Y_{ii} element of the admittance matrix. Recall that $y_s = jb_s$, and that the sign of b_s depends on the shunt being inductive or capacitive.

Knowing how to deal with shunts, it is now easy to incorporate the line shunt admittance model. In the admittance matrix, for a transmission line between the buses i and j, the halved line shunt admittance $\frac{y_s}{2}$ of that line, has to be added to both Y_{ii} and Y_{jj}. For a transmission line with shunt admittance y_s, we thus find

$$\begin{bmatrix} I_{ij} \\ I_{ji} \end{bmatrix} = \left(y_{ij} \begin{bmatrix} 1 & -1 \\ -1 & 1 \end{bmatrix} + y_s \begin{bmatrix} \frac{1}{2} & 0 \\ 0 & \frac{1}{2} \end{bmatrix} \right) \begin{bmatrix} V_i \\ V_j \end{bmatrix}. \tag{4.12}$$

The influence on the admittance matrix, of a device t between the buses i and j, that is either a tap transformer or a phase shifter, can be derived from the model. Let E be the voltage induced by t, then

$$V_i = TE. \tag{4.13}$$

Thus, the current from bus j to t in the direction of bus i is

$$I_{ji} = y_{ij} (V_j - E) = y_{ij} \left(V_j - \frac{V_i}{T} \right). \tag{4.14}$$

Conservation of power gives

$$V_i \overline{I}_{ij} = -E \overline{I}_{ji} \Rightarrow T \overline{I}_{ij} = -\overline{I}_{ji} \Rightarrow \overline{T} I_{ij} = -I_{ji}. \tag{4.15}$$

and thus, for the current from bus i to t in the direction of j we find

$$I_{ij} = -\frac{I_{ji}}{\overline{T}} = y_{ij} \left(\frac{V_i}{|T|^2} - \frac{V_j}{\overline{T}} \right). \tag{4.16}$$

If the device t that connects bus i to bus j is a tap transformer, then $\overline{T} = T$, and $|T|^2 = T^2$, and thus instead of the admittance matrix values from Eq. (4.12), we find

$$
\begin{bmatrix} I_{ij} \\ I_{ji} \end{bmatrix} = y_{ij} \begin{bmatrix} \frac{1}{T^2} & -\frac{1}{T} \\ -\frac{1}{T} & 1 \end{bmatrix} \begin{bmatrix} V_i \\ V_j \end{bmatrix}.
\tag{4.17}
$$

If instead t is a phase shifter, then $\overline{T} = e^{-j\delta_T} = \frac{1}{T}$, and $|T|^2 = 1$, and we find

$$
\begin{bmatrix} I_{ij} \\ I_{ji} \end{bmatrix} = y_{ij} \begin{bmatrix} 1 & -T \\ -\overline{T} & 1 \end{bmatrix} \begin{bmatrix} V_i \\ V_j \end{bmatrix}.
\tag{4.18}
$$

4.4 Load Flow

The load flow problem, or power flow problem, is the problem of computing the flow of electrical power in a power system in steady state. In practice, this amounts to calculate all node voltages and line currents in the power system. The load flow is regarded as the most important network computation. The problem arises in many applications in power system analysis, and is treated in many books on power systems, see for example [4] or [3].

The mathematical equations describing the load flow problem can be obtained by combining Ohm's law from Eq. (4.5), with Joule's law:

$$
S_i = V_i \overline{I}_i = V_i \left(\overline{YV} \right)_i = V_i \sum_{k=1}^{N} \overline{Y}_{ik} \overline{V}_k.
\tag{4.19}
$$

Note that the admittance matrix Y is easy to obtain, and generally very sparse. Therefore a formulation using the admittance matrix has preference over using the impedance matrix, which is generally a lot harder to obtain and not sparse.

Further note that for each bus i that has no injected power, i.e., $S_i = 0$, we also have that injected current $I_i = 0$. Therefore, we can use the linear equation $(YV)_i = 0$, to eliminate the variable V_i from the problem. This method is called Kron reduction, see [12] Sect. 9.3. The elimination can be done by simple Gaussian elimination, but in the case of many such buses, using a Schur complement may provide better results.

Recall that voltage and current are complex numbers, and substitute $V_i = |V_i| e^{j\delta_i}$, $Y = G + jB$, and $\delta_{ij} = \delta_i - \delta_j$ into Eq. (4.19). This gives

$$
S_i = |V_i| e^{j\delta_i} \sum_{k=1}^{N} (G_{ik} - jB_{ik}) |V_k| e^{-j\delta_k} = \sum_{k=1}^{N} |V_i| |V_k| (\cos \delta_{ik} + j \sin \delta_{ik}) (G_{ik} - jB_{ik})
\tag{4.20}
$$

Define the real vector of the voltage variables of the load flow problem as

$$
\mathbf{V} = [\, \delta_1, \ldots, \delta_N, |V_1|, \ldots, |V_N| \,]^T .
\tag{4.21}
$$

For the purpose of notational comfort, we further define

$$P_{ij}(\mathbf{V}) = |V_i||V_j| \left(G_{ij} \cos \delta_{ij} + B_{ij} \sin \delta_{ij} \right), \tag{4.22}$$

$$Q_{ij}(\mathbf{V}) = |V_i||V_j| \left(G_{ij} \sin \delta_{ij} - B_{ij} \cos \delta_{ij} \right). \tag{4.23}$$

Then we can write equation (4.20) as

$$S_i = \sum_{k=1}^{N} P_{ik}(\mathbf{V}) + j \sum_{k=1}^{N} Q_{ik}(\mathbf{V}). \tag{4.24}$$

Splitting the real and imaginary parts of the equation, and defining $P_i(\mathbf{V}) = \sum_{k=1}^{N} P_{ik}(\mathbf{V})$ and $Q_i(\mathbf{V}) = \sum_{k=1}^{N} Q_{ik}(\mathbf{V})$, we have

$$P_i = P_i(\mathbf{V}), \tag{4.25}$$

$$Q_i = Q_i(\mathbf{V}). \tag{4.26}$$

Equations (4.25)–(4.26) relate the complex power in each node to the node voltages, using the admittance matrix of the power system network. It consists of $2N$ non-linear real equations. Each node has four variables, $|V_i|$, δ_i, P_i and Q_i, two of which have a specified value, as discussed in Sect. 4.3. Thus we have $2N$ non-linear real equations in $2N$ real unknowns, commonly known as the load flow equations, or power flow equations.

4.4.1 Newton–Raphson

It is generally not possible to solve a system of non-linear equations analytically. However, there are many iterative techniques to approximate the solution. The problem of solving a system of equations is trivially equivalent to find the simultaneous roots of a set of functions in the same variables. As such, we can deploy root finding algorithms as a tool to solve a system of equations. Of these algorithms, Newton's method, also known as the Newton–Raphson method, is widely accepted to be the best for many applications. Newton's method is the base algorithm of choice when solving load flow equations.

First, define a real vector of the voltage variables of the load flow problem:

$$\mathbf{V} = [\,\delta_1, \dots, \delta_N, |V_1|, \dots, |V_N|\,]^T. \tag{4.27}$$

Then write the load flow equations as a set of functions $\mathcal{F}_i$, whos simultaneous roots are the solution of the load flow equations:

$$\mathcal{F}_i(\mathbf{V}) = \begin{bmatrix} \Delta P_i(\mathbf{V}) \\ \Delta Q_i(\mathbf{V}) \end{bmatrix} = \begin{cases} P_i - P_i(\mathbf{V}), & i = 1 \dots N, \\ Q_i - Q_i(\mathbf{V}), & i = N+1 \dots 2N, \end{cases} \tag{4.28}$$

where $S_i = P_i + jQ_i$ are the desired values of the power, whereas $S_i(\mathbf{V}) = P_i(\mathbf{V}) + jQ_i(\mathbf{V})$ are the values dictated by the bus voltages and network admittance matrix, i.e., $S_i(\mathbf{V})$ is the right-hand side in Eq. (4.20). The function $\mathcal{F}$ is called the power mismatch.

Next, we start with some guess or approximation $\mathbf{V}^0$, and update it iteratively by the rule

$$\mathbf{V}^{k+1} = \Phi\left(\mathbf{V}^k\right), \tag{4.29}$$

where the function Φ is such that

$$\Phi(\mathbf{V}) = \mathbf{V} \Leftrightarrow \mathcal{F}(\mathbf{V}) = \mathbf{0}, \tag{4.30}$$

with $\mathcal{F}$ the vector of functions $\mathcal{F}_i$. Condition (4.30) is always satisfied, if we choose

$$\Phi(\mathbf{V}) = \mathbf{V} - A(\mathbf{V})^{-1}\mathcal{F}(\mathbf{V}), \tag{4.31}$$

with $A(\mathbf{V})$ a non-singular matrix. Different choices of $A(\mathbf{V})$ correspond to different methods. For example, $A = I$ leads to the Gauss–Seidel method.

Newton's method is based on the first order Taylor expansion of $\mathcal{F}$, and as such uses $A(\mathbf{V}) = J(\mathbf{V})$, with J the Jacobian of $\mathcal{F}$. Thus, Newton's method is characterized by the update rule

$$\mathbf{V}^{k+1} = \mathbf{V}^k + \Delta\mathbf{V}^k, \tag{4.32}$$

where $\Delta\mathbf{V}^k$ is the solution of the linear system of equations

$$-J\left(\mathbf{V}^k\right)\Delta\mathbf{V}^k = \mathcal{F}\left(\mathbf{V}^k\right). \tag{4.33}$$

Convergence of Newton's method is not guaranteed. However, when the starting approximation is sufficiently close to the solution, the convergence is quadratic. Note that the Jacobian generally differs in each iteration. We will refer to this method as full Newton, as opposed to inexact Newton methods which make some kind of approximation, often in the Jacobian. Inexact Newton methods generally have linear convergence.

4.4.2 Generator Buses

When solving a linear system of the form $A\mathbf{x} = \mathbf{b}$, we assume the coefficient matrix A, and right-hand side vector $\mathbf{b}$ to be known, and the variable vector $\mathbf{x}$ to be unknown. In the case of the linear system from Eq. (4.33), each bus i of the power system makes for two equations, begin row i and row $N + i$. If bus i is a load bus, then indeed all coefficients, and the right-hand side value, of row i and row $N + i$ are known, whereas δ_i and $|V_i|$ are unknown. However, if bus i is a generator bus, then we have a different situation.

First let us consider the slack bus, of which every power system has exactly one, and assume that it is located at bus 1. Then, δ_1 and $|V_1|$ are known, whereas P_1 and Q_1 are not. As a result, we can eliminate the variables δ_1 and $|V_1|$ from the system, by substituting their values into the coefficient matrix, and bringing the resulting values to the right-hand side. Also, we can take row 1, which corresponds to the equation for P_1, and row $N + 1$, which corresponds to the equation for Q_1, out of the system. Note that the resulting coefficient matrix, is a principal minor of the original matrix.

Once the Newton–Raphson process has converged, P_1 and Q_1 are easily calculated by substituting the found solution into the original equations 1 and $N + 1$, that were taken out.

Now consider a generator bus i that is not the slack bus. Then P_i and $|V_i|$ are known, whereas Q_i and δ_i are unknown. With $|V_i|$ known, we can eliminate this variable, and column $N + i$ from the linear system. And since Q_i is unknown, we also take row $N + i$ out of the system, like we did in the case of a slack bus. Note that again, the resulting coefficient matrix is a principal minor of the original matrix.

4.4.3 Full Newton Jacobian

Applying Newton's method to the load flow problem, in each iteration we have to solve the linear system of equations (4.33), and update the approximate solution according to Eq. (4.32). All terms in these equations are straightforward to calculate, except for the Jacobian.

In this section we will derive the explicit form of the Jacobian of the load flow equations. The general structure of the Jacobian is

$$
J(\mathbf{V}) = - \left[\begin{array}{ccc|ccc}
\frac{\partial P_1}{\partial \delta_1} & \cdots & \frac{\partial P_1}{\partial \delta_{N_1}} & \frac{\partial P_1}{\partial |V_1|} & \cdots & \frac{\partial P_1}{\partial |V_{N_2}|} \\
\vdots & \ddots & \vdots & \vdots & \ddots & \vdots \\
\frac{\partial P_{N_1}}{\partial \delta_1} & \cdots & \frac{\partial P_{N_1}}{\partial \delta_{N_1}} & \frac{\partial P_{N_1}}{\partial |V_1|} & \cdots & \frac{\partial P_{N_1}}{\partial |V_{N_2}|} \\
\hline
\frac{\partial Q_1}{\partial \delta_1} & \cdots & \frac{\partial Q_1}{\partial \delta_{N_1}} & \frac{\partial Q_1}{\partial |V_1|} & \cdots & \frac{\partial Q_1}{\partial |V_{N_2}|} \\
\vdots & \ddots & \vdots & \vdots & \ddots & \vdots \\
\frac{\partial Q_{N_2}}{\partial \delta_1} & \cdots & \frac{\partial Q_{N_2}}{\partial \delta_{N_1}} & \frac{\partial Q_{N_2}}{\partial |V_1|} & \cdots & \frac{\partial Q_{N_2}}{\partial |V_{N_2}|}
\end{array} \right], \tag{4.34}
$$

where the dependance of P_i and Q_i of $\mathbf{V}$ has been left out for readability. The dimensions N_1 and N_2 are due to the elimination of the slack bus, and other generator buses.

Below the first order derivatives, of which the Jacobian exists, are derived. Note that we assume that $i \neq j$ wherever applicable.

$$\frac{\partial P_i\,(\mathbf{V})}{\partial \delta_j} = |V_i|\,|V_j|\left(G_{ij}\sin\delta_{ij} - B_{ij}\cos\delta_{ij}\right) = Q_{ij}\,(\mathbf{V}) \tag{4.35}$$

$$\frac{\partial P_i\,(\mathbf{V})}{\partial \delta_i} = \sum_{k\neq i} |V_i|\,|V_k|\left(-G_{ik}\sin\delta_{ik} + B_{ik}\cos\delta_{ik}\right) = -Q_i\,(\mathbf{V}) - |V_i|^2\,B_{ii}$$

$$\tag{4.36}$$

$$\frac{\partial Q_i\,(\mathbf{V})}{\partial \delta_j} = |V_i|\,|V_j|\left(-G_{ij}\cos\delta_{ij} - B_{ij}\sin\delta_{ij}\right) = -P_{ij}\,(\mathbf{V}) \tag{4.37}$$

$$\frac{\partial Q_i\,(\mathbf{V})}{\partial \delta_i} = \sum_{k\neq i} |V_i|\,|V_k|\left(G_{ik}\cos\delta_{ik} + B_{ik}\sin\delta_{ik}\right) = P_i\,(\mathbf{V}) - |V_i|^2\,G_{ii} \tag{4.38}$$

$$\frac{\partial P_i\,(\mathbf{V})}{\partial |V_j|} = |V_i|\left(G_{ij}\cos\delta_{ij} + B_{ij}\sin\delta_{ij}\right) = \frac{P_{ij}\,(\mathbf{V})}{|V_j|} \tag{4.39}$$

$$\frac{\partial P_i\,(\mathbf{V})}{\partial |V_i|} = 2|V_i|\,G_{ii} + \sum_{k\neq i} |V_k|\left(G_{ik}\cos\delta_{ik} + B_{ik}\sin\delta_{ik}\right) = \frac{P_i\,(\mathbf{V})}{|V_i|} + |V_i|\,G_{ii}$$

$$\tag{4.40}$$

$$\frac{\partial Q_i\,(\mathbf{V})}{\partial |V_j|} = |V_i|\left(G_{ij}\sin\delta_{ij} - B_{ij}\cos\delta_{ij}\right) = \frac{Q_{ij}\,(\mathbf{V})}{|V_j|} \tag{4.41}$$

$$\frac{\partial Q_i\,(\mathbf{V})}{\partial |V_i|} = -2|V_i|\,B_{ii} + \sum_{k\neq i} |V_k|\left(G_{ik}\sin\delta_{ik} - B_{ik}\cos\delta_{ik}\right) = \frac{Q_i\,(\mathbf{V})}{|V_i|} - |V_i|\,B_{ii}$$

$$\tag{4.42}$$

Summarizing, again leaving out the dependance on $\mathbf{V}$:

$$
\begin{array}{ll}
\dfrac{\partial P_i}{\partial \delta_j} = Q_{ij} & |V_j|\,\dfrac{\partial P_i}{\partial |V_j|} = P_{ij} \\[2mm]
\dfrac{\partial P_i}{\partial \delta_i} = -Q_i - |V_i|^2\,B_{ii} & |V_i|\,\dfrac{\partial P_i}{\partial |V_i|} = P_i + |V_i|^2\,G_{ii} \\[2mm]
\hline
\dfrac{\partial Q_i}{\partial \delta_j} = -P_{ij} & |V_j|\,\dfrac{\partial Q_i}{\partial |V_j|} = Q_{ij} \\[2mm]
\dfrac{\partial Q_i}{\partial \delta_i} = P_i - |V_i|^2\,G_{ii} & |V_i|\,\dfrac{\partial Q_i}{\partial |V_i|} = Q_i - |V_i|^2\,B_{ii}
\end{array}
$$

Algorithm 1 Newton–Raphson Method

1: $k := 0$
2: given initial iterate $\mathbf{V}_0$
3: **while** not convergenced **do**
4:     solve $-J(\mathbf{V}_k)\Delta\mathbf{V}_k = \mathcal{F}(\mathbf{V}_k)$
5:     update $\mathbf{V}_{k+1} := \mathbf{V}_k + \Delta\mathbf{V}_k$
6:     k := k+1
7: **end while**

4.4.4 Globalized Newton

The convergence of the Newton–Raphson method is known to critically depend on the choice of the initial guess $\mathbf{V}_0$. Convergence is likely to fail in case that the initial guess is too far from the solution. In load flow computations the initial guess is typically chosen as $\mathbf{V}_0 = \mathbf{1}$ (in per unit values). This is referred to as the *flat start* solution. With this initial guess, the Newton iteration is likely to converge in a few iterations in problems for which the solution $\mathbf{V}$ satisfies $|V_i| \approx 1$ and $\delta_i \approx 0$. In optimization studies and in Monte-Carlo simulations, the load flow simulation is placed in a outer loop [13]. For these problems the flat start is likely to lie outside the basin of attraction of the Newton method. The same is true for stressed networks for which the solution is likely to deviate substantially from $|V_i| \approx 1$ and $\delta_i \approx 0$. Some kind of save guarding from bad initial guesses is therefore mandatory.

To saveguard against bad initial guesses, a code that implements the Newton–Raphson algorithm is typically augmented with other code that iteratively improves the current initial guess using either line-search or trust-region algorithms. The update formula (4.32) is replaced by a trust-region subproblem. The Powell dogleg procedure [8] is an efficient method to solve this subproblem. Other globalization methods are discussed in [14].

To give an example of the effect of saveguarding against bad initial guesses, we give in Fig. 4.1 the convergence history of the Newton algorithm with and without saveguarding [15]. As example we took the test case **case2383wp.m** from the Mat-Power [11] test collection from to which we added a generator. Figure 4.1 shows the convergence of the Newton's method to solve the load flow equations with and without globalization. The algorithm without globalization is the implementation in the function `newtonpf.m` in MatPower package. The method with globalization using the Powell dogleg procedure is obtained using the function `fsolve` in Matlab. Without globalization the plot shows that Newton's method does not converge. With globalization instead convergence is obtained.

Algorithm 2 Inexact Newton–Raphson Method

1: $k := 0$
2: given initial iterate $\mathbf{V}_0$
3: **while** not convergenced **do**
4:     minimize $\mathcal{F}(\mathbf{V}_k + s\mathbf{W}_k)$
5:     update $\mathbf{V}_{k+1} := \mathbf{V}_k + \Delta\mathbf{V}_k$
6:     k := k+1
7: **end while**

4.5 Newton–Krylov Method

Newton's method requires solving a linear system at each iteration. In literature [3, 4] this linear system solve is typically performed by a direct solution method. Cur-

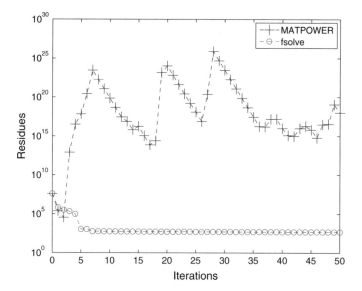

Fig. 4.1 Comparison of Newton-only (`newtonpf.m` in MatPower) and globalized Newton (`fsolve` in Matlab) algorithm. Plot of the residuals versus iterations for both Newton's method and the trust-region method for a version of **case2383wp.m** from MatPower [11] to which a generator was added. The converge of the trust-region method is seen to be more stable

rently however there exists a pressing need to compute the power flow in networks that are large in size. Such networks originate for instance in the modeling of the interconnection of the network Europe with that in Russia or in Northern Africa. It is well known that direct solution methods are not suited for such large scale problems [9, 16]. A new generation of load flow solvers in which the LU-factorization of Jacobian $J(\mathbf{V}_k)$ is replaced by a preconditioned Krylov subspace iteration is available in literature [2]. The term Newton–Krylov for these new solvers derives from the fact the accuracy of the inner linear solved is linked to the outer non-linear iteration residual. The PETSc software library is a public domain software library that implements a load flow solver called `pflow` [17] that allows to employ these modern iterative solvers. This solver employs the `DMNetwork` data structures to manage the network data, the System of Nonlinear Equation Solvers (`SNES`) component to handle the non-linearity of the equations and the Preconditioning (`PC`) and Krylov Subspace (`KSP`) components to provide the preconditioners and the Krylov subspace solvers.

In Table 4.1 we present statistics on the performance of the `pflow` solver applied to the test example **case6468rte.m** taken from MatPower [11]. As Krylov subspace solver we employ the Generalized Minimal Residual Method (GMRES). We vary the level of fill-in in the incomplete LU factorization from 0 to 16. We list the number of Newton iterations, the overall number of GMRES iterations (sum of 5 runs), the number of non-zero elements in the preconditioner, the level of fill-in in the preconditioner relative to the number of non-zero elements in the original

Table 4.1 Numerical results for the Newton–Krylov AC load flow method for the **case6468rte.m** test case from MatPower [11] for various level of fill-in in the ILU preconditioner without QMD reordering (top) and with QMD reordering (bottom). Tabulated are the number of Newton iteration, the total number of GMRES iterations, the number of non-zeros in the preconditioner, the fill-in ratio in the preconditioner and the total simulation time

Newton–Krylov AC load flow on **case6468rte.m**

Without QMD reordering of J ($\mathbf{V}_k$)

Preconditioner	ILU(0)	ILU(2)	ILU(4)	ILU(8)	ILU(16)
Newton iterations	–	5	5	5	5
GMRES iterations	–	541	152	73	20
nnz(L+U)	–	215.648	413.080	1.449.264	7.063.408
Fill ratio	–	2.39	4.57	16.03	78.14
Time(s)	–	2.133	1.930	4.658	77.58

With QMD reordering of J ($\mathbf{V}_k$)

Preconditioner	ILU(0)	ILU(2)	ILU(4)	ILU(8)	ILU(16)
Newton iterations	5	5	5	5	5
GMRES iterations	1213	163	76	40	15
nnz(L+U)	90.392	115.104	126.520	138.600	1.495.584
Fill ratio	1	1.27	1.49	1.53	1.65
Time(s)	3.025	1.277	1.092	1.044	0.988

matrix and the overall CPU time. We compare runs without (top) and with (bottom) use of Quotient Minimal Degree reordering. In all runs, 5 Newton iterations are required to reach convergence. Both with and without QMD reordering, the number of GMRES iterations decreases as expected as the level of fill-in in the preconditioner is increased. Without the use of QMD reordering, however, the number of non-zeros in the preconditioner, the level of fill-in in the preconditioner and the overall CPU time increases significantly. With the use of QMR reordering in contrast, the fill ratio of the preconditioner remains bounded by 1.65. When QMD reordering is used, the highest level of fill-in in the preconditioner is seen to deliver the shortest overall run time. More results on the Newton–Krylov method for load flow computations are available in [2, 18].

4.6 Newton–Krylov–Schwarz Method

In particular applications, it is useful to view a power system network as an interconnection of various subnetworks. An illustrative example is given in Fig. 4.2. This figure shows the decomposition of the **case9.m** test case from MatPower into subnetworks. Subnetworks can be owned and operated by different entities. Despite the shared ownership, one would like to simulate the entire network without violating sensitive issues on the propriety of the network data. This problem is referred to as co-simulation and is discussed extensively elsewhere in this book.

The Newton–Krylov methodology discussed in this chapter can be adapted to allow some form of co-simulation of an interconnection of various power systems. Indeed, by adopting the Schwarz method as a preconditioner for the linear system solve at each Newton iteration, the linear solver can exploit the network decomposition. The Schwarz method was developed as a domain decomposition method for solving large scale elliptic and parabolic partial differential equations in parallel [19, 20]. In its most simple form, the Schwarz preconditioner is equivalent to be block Jacobi (additive form) or block Gauss–Seidel (multiplicative form). Schwarz methods decompose the problem into subproblems, perform local solves and employ interface conditions to stitch the individual parts together to obtain the global solution.

The load flow solver `pflow` implemented in PETSc [10] does provide Schwarz-type preconditioners. This solver can therefore be used to solve the load flow equation in a distributed fashion. We applied the Newton–Krylov–Schwarz algorithm implemented implemented in `pflow` to a synthetically constructed test case that we will refer to as test case **case5timespegase.m**. This test case was constructed by doubling the test case **case9241pegase.m** from MatPower five times and establishing links between these copies. It resulting network has $9.242 * 32 = 295.712$ nodes. The procedure followed to construct the test case is the same as the one outlined in [2]. Numerical results are given in Table 4.2. This table lists the number of Newton iterations, the total amount of linear iterations and the total CPU time for a decomposition into two (top) and four (bottom) subnetworks. For both decompositions, we tested an additive form of the Schwarz method for various amounts of overlapping nodes ranging between 0 and 2. The results in Table 4.2 show that the number of Newton iterations does not vary with the number of subnetworks and the amount of overlap. This can be explained by the fact that, at each Newton iteration, the linear system is solved to full accuracy for all parameter settings. The number of linear iterations decreases significantly by increasing the overlap from 0 to 1. Increasing the amount of overlap even more does not result in a significant further reduction in the number of iterations. The number of linear iterations increases with the number of subnetworks. More iterations are indeed required to pass information from one network to all the others. This report on the number of iterations is similar to what is reported in the literature on the Schwarz method for discretized elliptic partial differential equations, see [19, 20] and the references cited therein. The CPU time is a complex function of the number of iterations, the amount of overlap and the number of subnetworks. More results of distributed load flow computations are available in [21].

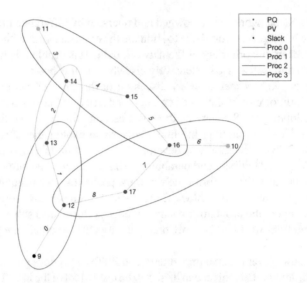

Fig. 4.2 Decomposition of the **case9.m** network into four subnetworks

Table 4.2 Numerical results for the Newton–Krylov–Schwarz AC load flow method applied to the **case5timespegase.m** test case. This test case was constructed by copying **case9241pegase.m** from MatPower [11] five times and interconnecting these copies. Decompositions into two and four subnetworks with various amounts of overlapping nodes ranging between 0 and 3 were tested. Tabulated are the number of Newton iteration, the total number of GMRES iterations and the total simulation time

Newton–Krylov–Schwarz AC Load Flow on **case5timespegase.m**			
Decomposition in two subnetworks			
Overlapping edges	Non-linear its.	Linear its.	CPU time (s)
0	7	72	81
1	7	38	77
2	7	30	72
3	7	30	76
Decomposition in four subnetworks			
Overlapping edges	Non-linear its.	Linear its.	CPU time (s)
0	7	100	54
1	7	49	51
2	7	39	50
3	7	37	62

4.7 Conclusions

We discussed the globalized Newton–Krylov–Schwarz method to solve the AC load flow equation in power systems. We showed that the globalization of the Newton method is required to obtain convergence for stressed networks. We argued that Newton–Krylov methods yield efficient solvers and that Schwarz methods allow to distribute computations over subnetworks.

Acknowledgements The results on the globalization of the Newton method using the `fsolve` function in Matlab resulted from a fruitful collaboration with VVTP Applied Physics student association. Results from the Newton–Krylov method using `pflow` implemented in PETSc resulted from the master thesis project of Jonathan Aviles. Results from the Newton–Krylov–Schwarz using again `pflow` resulted from the master thesis project of the students Andrea Ceresoli and Stefano Guido Rinaldo.

References

1. J. Arrillaga, C.P. Arnold, *Computer Analysis of Power Systems* (Wiley, New Jersey, 1990)
2. R. Idema, D. Lahaye, *Computational Methods in Power System Analysis* (Atlantis Press, Amsterdam, 2014)
3. W.H. Kersting, *Distribution System Modeling and Analysis*, 3rd edn. (Taylor & Francis, Abingdon, 2012)
4. P. Schavemaker, L. van der Sluis, *Electrical Power System Essentials* (Wiley, New Jersey, 2008)
5. M.H. Bollen, F. Hassan, *Integration of Distributed Generation in the Power System*, IEEE Press Series on Power Engineering (Wiley, New Jersey, 2011)
6. N. Hatziargyriou, *Microgrids: Architectures and Control* (IEEE, Wiley, New Jersey, 2014)
7. http://www.openelectrical.org/wiki/index.php?title=Power_Systems_Analysis_Software
8. J. Nocedal, S. Wright, *Numerical Optimization* (Springer, New York, 2006)
9. Y. Saad, *Iterative Methods for Sparse Linear Systems*, 2nd edn. (SIAM, Philadelphia, 2003)
10. S. Balay, S. Abhyankar, M.F. Adams, J. Brown, P. Brune, K. Buschelman, L. Dalcin, V. Eijkhout, W.D. Gropp, D. Kaushik, M.G. Knepley, L.C. McInnes, K. Rupp, B.F. Smith, S. Zampini, H. Zhang, H. Zhang, PETSc Web Page (2016), http://www.mcs.anl.gov/petsc
11. C.E. Murillo-Sánchez, R.D. Zimmerman, C.L. Anderson, R.J. Thomas, Secure planning and operations of systems with stochastic sources, energy storage and active demand. IEEE Trans. Smart Grid **4**(4), 2220–2229 (2013)
12. A.R. Bergen, V. Vittal, *Power Systems Analysis* (Pearson/Prentice Hall, New Jersey, 2000)
13. M. de Jong, G. Papaeffhymiou, D. Lahaye, C. Vuik, L. van der Sluis, Impact of correlated infeeds on risk-based power system security assessment, in *Power Systems Computation Conference (PSCC)* (Wroclaw, Poland, 2014). https://doi.org/10.1109/PSCC.2014.7038439
14. P.J. Lagacé, M.H. Vuong and I. Kamwa, Improving power flow convergence by Newton Raphson with a Levenberg-Marquardt method, in *IEEE Power and Energy Society General Meeting-Conversion and Delivery of Electrical Energy in the 21st Century* (2008), pp. 1–6
15. M. De Beurs, P. De Graaf, P. Hansler, S. Hermans, K. Van Walstijn, J. De Winter, D.J.P. Lahaye, Optimal configuration of the future electricity grid. DIAM TU Delft Technical Report 16-01 (2016)
16. R. Idema, G. Papaefthymiou, D. Lahaye, C. Vuik, L. van der Sluis, Towards faster solution of large power flow problems. IEEE Trans. Power Syst. **28**(4), 4918–4925 (2013)
17. S. Abhyankar, B.F. Smith, PETSc: an advanced math and computing framework for rapidly developing parallel smart grid applications, in *Proceedings of the IEEE PES General Meeting* (2013)

18. J. Aviles Cedeño, A three-phase unbalanced load flow solver for large-scale distribution power systems. TU Delft Master thesis (2017). (uuid:0d750fa1-b349-4459-8ba7-5f2a3bbf0c87)
19. A. Toselli, O. Widlund, *Domain Decomposition Methods - Algorithms and Theory*, Springer Series in Computational Mathematics (Springer, Berlin, 2004)
20. B. Smith, P. Bjorstad, W. Gropp, *Domain Decomposition: Parallel Multilevel Methods for Elliptic Partial Differential Equations* (Cambridge University Press, Cambridge, 2004)
21. S. Guido Rimaldo, A. Ceresoli, Newton-Krylov-Schwarz methods for distributed load flow and related applications. Master thesis report, School of Industrial and Information Engineering, Politecnico di Milano (2018)

Chapter 5
Co-simulation of Intelligent Power Systems

**Claudio David López, Miloš Cvetković, Arjen van der Meer
and Peter Palensky**

Abstract The complexity of energy systems increases as more renewable generation and energy storage technologies are added to the grid. Diverse energy carriers are becoming interconnected and the grids are getting reliant on communication networks for timely operation. The arising complexity is difficult to model with the existing mathematical models and using existing simulation tools due to confinement of these models and tools to a subset of the interconnected system. To overcome this challenge, combined simulation (co-simulation) methodology is being deployed. In co-simulation, multiple models and tools are being interconnected to truthfully represent reality. In this work, we review several aspects of co-simulation. First, we look at interconnecting transmission and distribution grid simulations in order to enable collaboration between transmission system operators (TSOs) and distribution system operators (DSOs). Next, we investigate co-simulation as means to dynamic model exchange between TSOs. Finally, we analyze co-simulation capabilities for running experiments in remotely connected research labs.

5.1 Coupled Simulation of Multiarea and Transmission/Distribution Systems

The unprecedented complexity of modern power system has called into question the traditional approach to their analysis. In this traditional approach, an area of interest in the system is selected and analyzed in detail, while its surrounding areas are

C. D. López (✉) · M. Cvetković · A. van der Meer · P. Palensky
Faculty of Electrical Engineering, Mathematics and Computer Science,
Delft University of Technology, Delft, The Netherlands
e-mail: c.d.lopez@tudelft.nl

M. Cvetković
e-mail: m.cvetkovic@tudelft.nl

A. van der Meer
e-mail: a.a.vandermeer@tudelft.nl

P. Palensky
e-mail: p.palensky@tudelft.nl

© Springer Nature Switzerland AG 2019
P. Palensky et al. (eds.), *Intelligent Integrated Energy Systems*,
https://doi.org/10.1007/978-3-030-00057-8_5

represented by simplified equivalent models. For example, at the transmission level a distribution grid could be represented as a current source and at the distribution level a transmission grid could be represented as a voltage source or a generator with a large inertia constant. While the use of equivalent models certainly simplifies the analysis, technologies like power electronics and ICT have only made it more difficult to come up with equivalents that are truly representative of the grids they stand for.

As the diversity and complexity of the technologies that can be found in a power system increases, so does the need for more detailed analysis methods. These methods should properly account for the uniqueness of each area in the system and the interactions between neighboring areas. Two main approaches to address this need follow: developing better equivalent models and using detailed models instead of their equivalents.

Equivalent models have the inherent advantage of reduced computational burden when compared to detailed models. However, the right assumptions need to be made to ensure representativity. Furthermore, as systems become more complex, the complexity of the methods required for obtaining the equivalent model also increases. One way to eliminate the challenge of creating equivalent models that are indeed representative would be to use full models, but there are some practical obstacles in this case as well.

Oftentimes a full model of a grid is not available, either because it does not exist or because it is owned by a third party that, out of privacy concerns, is unwilling to share it. A similar challenge can be encountered if instead the model is to be developed from scratch; the grid data required to develop the model might be owned by a third party and it might be confidential. Furthermore, even if a full model is already available, simulating a multiarea system in such detail would be very computationally expensive. Additionally, such a large model would be highly labor intensive to maintain.

5.1.1 Co-simulation as an Approach

In a co-simulation several independent simulators, each simulating only a part of a larger system, collaborate at runtime by exchanging data. The data exchange and the synchronization of the local simulation time of each simulator is orchestrated by a so called co-simulation master. Since the data exchange can be over a communication network, the simulators can be geographically distributed. This open up the possibility for different institutions to simulate collaboratively while bypassing any limitations due to confidentiality of information and/or models. Using co-simulation for multiarea system analysis has a set of additional advantages:

- Access to data: The model of each area can be developed independently by the institution that has access to the information needed.

- Tool flexibility: The simulation tool of choice for each area is irrelevant since data is exchanged over a network using a standard protocol.
- Extensibility: New simulators or models can be easily coupled.
- Privacy: The models can remain private if required since only selected simulation variables need to be shared with other simulators at run time.
- Reduced work load: Tasks related to model development and model maintenance are naturally divided among those that have access to grid information.
- Reduced computational load: The distributed nature of the co-simulation allows for the computational load of the co-simulation to be shared.

5.1.2 Challenges

Implementing such a co-simulation does not come without its challenges. The first challenge that must be overcome stems from the need to accommodate closed-source simulators in the co-simulation. In many cases this means that the co-simulation interfaces needed to couple the simulators in a co-simulation environment lack certain functionalities, for example, time roll back [1]. In many cases it becomes necessary to resort to hacks in order to create the interfaces. Another challenge, also related to closed-source simulators, has to do with step size control. The easiest case to manage is when all simulators use the same time step size and this remains constant during the co-simulation. However, there are cases when this cannot be ensured. In such cases the complexity of the synchronization mechanisms increases rapidly, especially when a potentially large number of simulators participate in the co-simulation. Numerical accuracy and stability of co-simulations can be difficult to ensure as well, and the limited control that users have over a simulator often stands in the way of certain co-simulation methods that are useful for addressing numerical challenges [2]. Managing a co-simulation like this is also a challenge, as a large number of simulators can be involved. Tools like mosaik [3] have been developed to address this problem at the software level, but the management challenge goes beyond software when several stakeholders are involved in running a co-simulation collaboratively. The performance of these co-simulations is a concern as well, especially if the number of involved simulators is large or if the frequency of data exchange is high, which is the case of EMT-type co-simulations [4].

5.1.3 Environment

The co-simulation environment developed to research this type of co-simulation is depicted in Fig. 5.1. The environment is composed of a set of Windows Server 2012 virtual servers, each running PowerFactory 15, plus an additional virtual server running the co-simulation master. The co-simulation master was developed in Python specifically for this application. The simulators and the master exchange

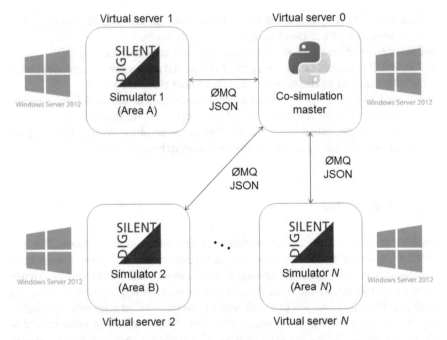

Fig. 5.1 Co-simulation environment for dynamic multiarea power system analysis

JSON-encoded messages over ØMQ sockets. These messages contain simulator inputs/outputs and all the necessary information for time synchronization.

5.1.4 Example

As an example of a multiarea co-simulation run with the previously described co-simulation environment, let us consider the transmission grid from Fig. 5.2 and the distribution grid from Fig. 5.3. The distribution grid is connected to Bus 6 of the transmission grid.

Figure. 5.4 shows the co-simulation environment and the results of a co-simulation of the two systems from Figs. 5.2 and 5.3. In this co-simulation a three-phase to ground fault occurs on the low voltage side of the distribution transformer (Fig. 5.3) at 0.05 s. Figure 5.5 shows some of the co-simulation results in more detail. This figure compares the phase voltage of one phase as measured on each side of the co-simulation interface (v_a). The same comparison is made with the current flowing through the interface (i_a) (between the transmission and distribution grids). There is almost no perceptible difference between voltages and currents on each side of the interface, which might give the false impression of high co-simulation accuracy. However, the difference on the power flowing through the co-simulation interface

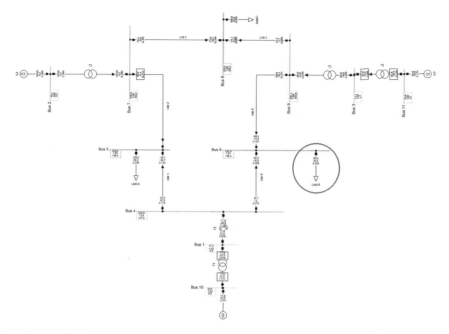

Fig. 5.2 IEEE 9 bus transmission system. The load marked in red represents the distribution grid. This grid is modeled as an ideal current source

(p_a), also shown in Fig. 5.5, is much more pronounced. This means that the power that the transmission grid is sending to the distribution grid is not the same as the power the distribution grids receives from the transmission grid, which is a violation of the law of conservation of energy. The difference between the power flow on each side of the co-simulation interface (Δp) provides a measure of how inaccurate the co-simulation actually is. In this case, while the co-simulation is in steady state the Δp is very small, but as soon as a disturbance occurs in the system, the Δp reaches a value beyond 2 MW. Despite the loss of accuracy that a co-simulation might display, its advantages still stand. Engineering judgment is required to determine whether these advantages outweigh the loss of accuracy or if more sophisticated co-simulation methods that have higher computational burden but deliver more accurate results are needed, for example, iterative methods [1].

In this section, we introduced co-simulation, its advantages and disadvantages, and we illustrated its accuracy in a representative use case. In the following section, we compare co-simulation to other methods for exchange of dynamic models between TSOs.

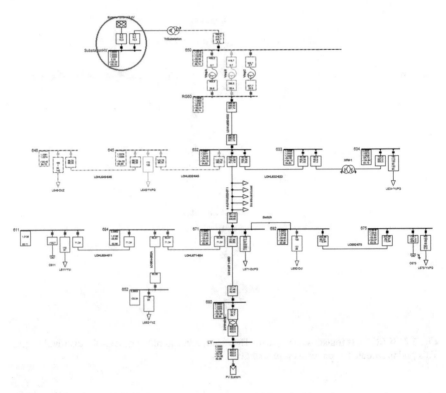

Fig. 5.3 IEEE 13 bus distribution system. The external grid marked in red represents the transmission grid. This grid is modeled as an ideal voltage source

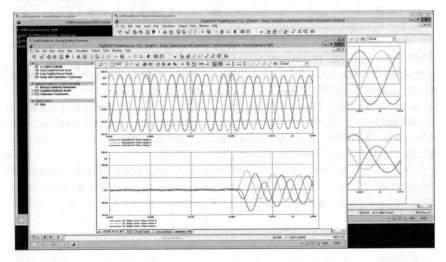

Fig. 5.4 Co-simulation environment running PowerFactory in two servers and an additional server for the co-simulation master. Some co-simulation results can be observed

Fig. 5.5 Phase voltage v_a, phase current i_a and phase instantaneous power p_a at and through the interface as measured on the transmission and distribution sides of the co-simulation. The difference between the power flowing through each phase of the interface as measured on the transmission and distribution (Δp) side provides a measure of the inaccuracy of the co-simulation

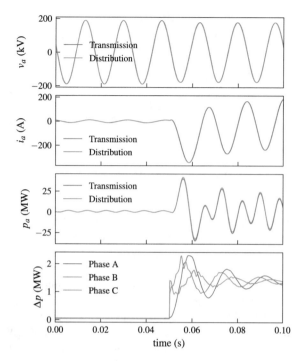

5.2 Dynamic Model Exchange and Co-simulation

Data and model exchange between TSOs is streamlined in practice using Common Grid Model Exchange Standard (CGMES) [5]. At this point of its development, version 2.5, one of the key features targeted for improvement is model exchange for dynamic studies, particularly in terms of compatibility with user-defined models of novel controllers and prototypes of new equipment.

In addition to the targeted improvements, multiple possibilities for the choice of the toolchain for dynamic model exchange are being considered. The main technical challenge for sharing the models is that TSOs keep and maintain the models within commercial simulation tools of their choice. These models are developed and updated over a span of many years, sometimes even decades, on the occasions of system expansion and component model validation. Since different vendors supply different TSOs, the problem of finding the appropriate means for model exchange becomes the problem of finding an adequate technological solution for sharing the information between simulation tools.

The possibilities to address this challenge are diverse (see Fig. 5.6). The simplest approach that comes to mind is to share a static data format, similarly to already established practice with static power flow models (see CGMES rules on static model exchange [5]). There are two main benefits of using such approach. First, the required technical solutions are relatively simple to design and implement.

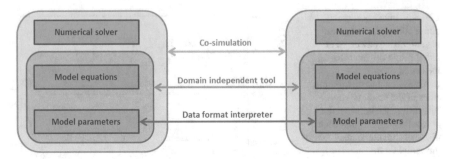

Fig. 5.6 Approaches to dynamic model exchange

Data format interpreters that are capable of exporting model information into the desired format and importing it back from the same format are easy to create using well established parser definitions. Second, the TSOs have been historically open to exchanging information in terms of static datasets [6]. Hence, the adoption barrier is relatively small. However, the accompanying challenges of this approach are far from straightforward to address. The main difficulty is that different simulation tools often use different component models that are not always comparable. For example, one tool might support the model of a synchronous generator with a single damper winding while the other tool might not. In such a case, the time-domain responses of the two simulators would yield different results. To overcome this obstacle, the data format interpreter would need to have a sufficient level of intelligence to handle these model misalignments in a most adequate manner. Defining the KPIs that describe the *most adequate manner* is a challenging task that would require considerable attention.

A slightly more involved approach to model exchange is to use a single target platform (i.e. simulation tool) when exporting the models [7]. If all TSOs export their models to the same target platform, sharing the models among them becomes trivial. The most appropriate choice for a target platform seems to be a domain independent modeling and simulation tool. The domain independent modeling and simulation tools, such as Modelica and Matlab-Simulink, are created for addressing wide range of applications. Thus, they use basic mathematical operations and functions as elementary building blocks (e.g. Modelica uses differential and algebraic equations in acausal manner while Simulink uses basic mathematical operations in causal manner including differentiation and integration and user-defined script functions [8]). In theory, since power system components are physical hardware components, any model encapsulated within proprietary power system simulation tools can be represented in a domain independent tool. Thus, the models can be exported and then simulated in a domain independent tool. This process is referred to as *model migration*. Although exporting models is generally more involved than exporting data, this task is deemed feasible if/when the TSOs develop trust and conformity with the domain independent tool in question.

However, one of the main technical challenges still remains. The commercial tools for simulation of power system dynamics often deploy proprietary modifications of numerical solvers in order to improve the performance of the tool and keep the competitive edge. At times, the proprietary modifications are also applied to the models. If/when this happens, no guarantees exist that the domain independent tool and the proprietary software tool will yield the same time domain response.

Simulations of power electronically enriched systems pose additional challenges for domain independent tool approach. The manufacturers of converters often keep their models encapsulated as black box models. Such models would be impossible to export to a domain independent modeling and simulation tool without the involvement of the vendor. In addition, some domain independent tools, such as Modelica, are created with open source policy in mind and their support for protecting proprietary models is limited (although possible using Functional Mock-up Units - FMUs [9]), making this approach to model exchange more involved.

Another difficulty with using domain independent tool appears in simulations of any large expansion project. Each large expansion project comes with specific design requirements and components that are often unique for that particular project. In such a case, user defined models must be created to capture the relevant dynamics. Migrating user defined models to a domain independent tool represents a challenge and an open problem [5]. PowerFactory for example uses its own proprietary scripting language (DSL) and its own proprietary block-diagram modeling tool (DPL). A user could, in all their freedom, develop a component so complex, that the migration from PowerFactory to Modelica becomes extremely difficult. The same challenge appears in the case of heterogeneous components at the distribution system level and in simulations of new technologies.

Finally, the co-simulation can be considered as the third and most comprehensive approach to dynamic model exchange. In fact, it is more appropriate to consider this approach as an alternative to model exchange, since it does not involve any model exchange per se. Co-simulation is created by exchanging the data between the proprietary simulators in run-time. Thus, there is no need to export the models and the parameters. Instead, the values of the so-called coupling variables are being exchanged as the simulations are executed. For example, if two TSOs wish to run transient stability simulations, they would couple their simulation tools. Each simulation tool would run the models and solvers that it typically runs with one addition, the boundary condition information would be exchanged between the tools in run-time. These boundary condition information are also referred to as exogenous information for each of the tools, or coupling variables on the interface between the tools. This approach guarantees that even the vendor-made adjustments to the model-solver combination together with the complex user-defined models would be accurately included in the resulting simulation response. At the same time, the TSOs would have high confidence in correctness of the response.

It should be noted that verification of the models is a challenge that accompanies all three of these approaches, although to a slightly different extent and in a different form. When using a data format interpreter or domain independent tool, the burden of verification is placed on the TSO that is exporting the models. The TSO would have to ensure that the imported model produces the same response as the model that is being exported. Since the TSO has the complete trust in its operating models, the verification would typically be done against the original model and the original simulation tool. In the case of the data format interpreter, this might have to be done for many different target platforms (PowerFactory, PSS/E, PSS/Netomac, PSCAD, etc.) which significantly increases the effort and expertise of the involved personnel. In the case of a specific domain independent modeling tool, the effort would be reduced to a single tool.

In the case of co-simulation, verification of time responses is more difficult to perform. Two alternatives exist: verification against a simulation tool, and verification against the field data. If verification is done against another (commercial) simulation tool, one would have to create exactly the same reference model in this commercial simulation tool since such reference model does not exist. In doing this, one would repeat the same work that would have to be done if data format interpreter or domain independent modeling tool are used for model exchange. Thus, this verification process defeats the purpose. Verification using field data is another option. Since existing TSO models are anyhow verified using signatures from the field data, extending this approach to verify co-simulation seems feasible in practice (albeit not without difficulties since the data belongs to several organizations).

In the rest of this section, we compare model migration to a domain independent modeling and simulation tool (Modelica in particular) with the co-simulation approach for dynamic model exchange. First, we introduce the case study used for comparison. Next, we analyze the results and identify similarities and differences between the obtained time responses. Finally, we outline possible technical reasons for the discrepancy between the signatures.

Case Study: Comparison of Model Migration and Co-simulation

The case study for comparison is chosen with several requirements in mind. First, the case study must be simple enough to minimize possible factors for potential discrepancies between two approaches. Yet, it has to provide sufficiently accurate representation of typical power system dynamics. Second, commonly used and industry adopted tool must be deployed as a benchmark for verification purposes.

To satisfy these two requirements we choose the same case study as in [10]. In this case study, the power system model is a three bus system. A synchronous generator is connected to Bus 1 and two identical loads are connected to Buses 2 and 3. One transmission line connects Buses 1 and 2 and another one connects Buses 1 and 3.

The benchmark simulation is conducted in PowerFactory 15.2. To illustrate model migration, the same power system has been modeled in Modelica using models developed in [7]. To illustrate co-simulation, the PowerFactory model is divided into two. The load at Bus 3 and transmission line connecting buses 1 and 3 are simulated in one instance of PowerFactory while the rest of the system runs in another instance of PowerFactory. The exchange of data among the simulators and synchronization of simulation execution is achieved by using a light weight co-simulation master algorithm from [4].

It is important to mention one detail regarding the following comparison. Model migration is benchmarked against Root Mean Square (RMS) simulation while co-simulation is benchmarked against Electromagnetic Transient (EMT) simulation. This is a non-ideal situation since the use of two different modeling paradigms allows only indirect comparison of model migration and co-simulation. The only reason to take this approach is of practical nature. Currently, model migration of RMS models is much easier to perform. At the same time, co-simulating RMS models is much more difficult than co-simulating EMT models. Thus, the difference in the two taken approaches.

The comparison is performed under two characteristic events. The first simulated event is a short circuit at Bus 2 in duration of 0.2 s created after 5 s of simulation time. The second event is a complete loss of load at Bus 2 that occurs at the 5 s mark and lasts until the end of simulation. Even though the simulations are performed from 0 to 10 s, only the most interesting time period of simulation is shown in the plots that follow.

We compare percentage difference error for some characteristic variables of this test case. We look at electrical torque and rotational speed of the synchronous generator and voltage at the generator terminal (Bus 1). The percentage difference error is computed according to the following equation

$$err(t) = \frac{x(t) - \bar{x}(t)}{\bar{x}(t)} \cdot 100\% \qquad (5.1)$$

where $x(t)$ is the variable under scrutiny obtained as a result of model migration or co-simulation while $\bar{x}(t)$ is the same variable obtained from the benchmark simulation.

We observe the results of the simulation runs in Fig. 5.7. By inspection, we see that the error is always the largest at the moment of event. This is expected since the switching events excite dynamics on all time scales and are sometimes handled differently by different tools. With all but one studied variable, the size of error is within reasonable range. The only variable with the higher error value, terminal voltage in the short circuit case in Fig. 5.7f, is the one that is most sensitive to the events in the grid. The high error value can be explained by the small values of benchmark voltage during the short circuit event (see Eq. (5.1)).

We also observe that co-simulation is more sensitive to events than model migration (see Fig. 5.7e). This is in part due to the modeling differences in RMS and EMT (EMT is more detailed, and thus, more sensitive to the events). The second reason, which is of higher interest for this work, is that the co-simulation master is less optimized to handle events than the internal solver of a monolithic simulation tool. For example, our co-simulation master does not have access to the system Jacobian while the internal solver does. This is a characteristic of the co-simulation approach that generally results in less accurate responses immediately after the event.

Another difference between model migration and co-simulation is that error in the case of model migration typically takes longer to settle (see Fig. 5.7d). This is mostly attributed to the differences between RMS and EMT. At the same time, the value of error in steady-state is sometimes larger in the case of co-simulation (see

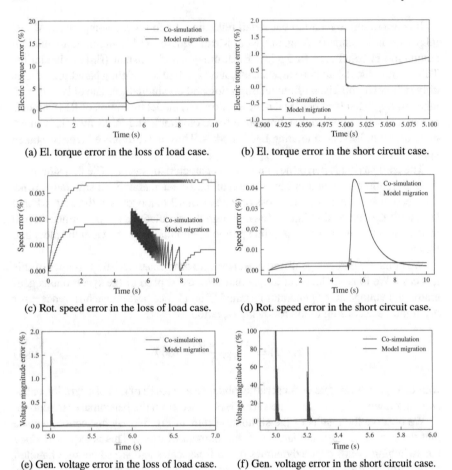

(a) El. torque error in the loss of load case.

(b) El. torque error in the short circuit case.

(c) Rot. speed error in the loss of load case.

(d) Rot. speed error in the short circuit case.

(e) Gen. voltage error in the loss of load case.

(f) Gen. voltage error in the short circuit case.

Fig. 5.7 Comparison of errors in model migration and co-simulation

Fig. 5.7a) and sometimes in the case of model migration (see Fig. 5.7b). The steady-state error depends on the models and also on the initialization approach and the choice of interfaces for co-simulation. These are different from one tool to another and can, to some extent, impact the dynamic response as well.

In this section, we compared two approaches for dynamic model exchange, model migration and co-simulation. As illustrated in the case study, both approaches have potential to achieve high accuracy. In the next section, we look at the role of co-simulation in the holistic testing and validation of smart grid experiments.

5.3   Holistic Testing and Validation of Cyber-Physical Energy Systems

Assets in power systems commonly have a life span of decades, being the primary equipment such as transformers, cables, and switch gear, or the secondary equipment like protective relaying and communication infrastructure. Such high durability puts a burden on the associated control algorithms and the energy management system, which as a platform have to maintain semantic compatibility across devices and subsystems for a long time. This used to hamper the deployment of innovative concepts and the operation of the power system is therefore very conventional (centralised control based on the physical properties of the connected units).

Successive technological developments have led to more rapid deployment of new primary and secondary equipment. For instance, over the past decade TSOs deployed well-controllable power electronics at high voltage levels, potentially offering devices the capability to behave as conventional power plants at the grid interface. This makes the nature of the primary source of a lesser concern and fosters the coupling of multi-energy to the electricity grid. The rapid digitalisation of our society on the other hand enables faster, widespread communication, and massive data acquisition, which is not left unnoticed in the power system. Decentralised control, automatic coordinated control, fast supervisory interventions by a centralised entity are striking examples that are current practice, i.e., smart grids. Above all, this makes interactions inside the power system faster and largely based on controls and its associated ICT infrastructure rather than physical response.

The entanglement of the multi-energy power system with ICT infrastructure leads to an unprecedented level of heterogeneity: the cyber-physical energy system (CPES). To maintain the same level of comfort and reliance on the electrical energy in our lives, measures for ensuring reliability and security of supply must be followed in the CPES. It is therefore significant to chart how

- domains such as electricity, ICT, and heat interact with each other, and
- test and validate new concepts for smart grids.

Testing new concepts is a challenge because laboratories are traditionally specialised into one particular domain such as ICT security, high-voltage electrical equipment, high-power electrical devices, etc. Moreover, a categorisation can be made in terms of experimental focus, such as laboratories focusing on pure hardware experiments, pure software experiments, or a combination like hardware in the loop (HIL) assessment. Testing a smart grid concept hence commonly yields drastic simplifications of the interconnected domains or a non-ideal experimental setup. For the conventional power system this was not an issue and abstracting out the boundaries of the system was common practice. Nowadays, the heterogeneity is more prominent and holistic system validation spanning multiple domains and utilising various experimental concepts is considered paramount.

The complexity of the CPES hence calls for a system-wide, cross-disciplinary procedure to test, validate, and roll-out smart grid concepts. Under the umbrella of the ERIGrid research project various smart-grid research infrastructures across Europe have set out a formal testing and validation procedure to cover these needs [11]. It combines the merits of established (quasi-)standards (e.g., smart grid architecture model, common information model, and unified modelling language [12]) with the testing expertise of laboratories. Besides subsystem-level validation the procedure aims to make tests transferable (and even partitionable) among research infrastructures and facilitate reproduction of experimental results.

5.3.1 Holistic Testing Procedure

In [13] the concept of *holistic testing* was introduced as being *"The process and methodology for evaluation of a concrete function, system or component within its relevant operational context with reference to a given test objective."*. Especially the system and their components exhibit functions that are cross-domain. Take as an example a centralised controller inside a wind power plant: physical quantities like voltages and currents are measured at the terminals of the wind turbines and transported to the park controller across dedicated communication channels over for instance IEC61850. At the connection point of the wind park the voltage and reactive power exchange with the transmission system shall be maintained within strict boundaries. This can be achieved by dispatching reactive power setpoints to the individual wind turbines. The communication channels can, however, cause latencies inhibiting the operation of the wind park as such. This cross-coupling of the physical, control, and ICT domain is very prominent here and needs to be carefully considered during specialized component and system testing.

To facilitate a bit of structure in such complex systems it is therefore important to specify which components and systems interact and how this relates to actual test objective. In ERIGrid this led to the holistic testing approach, first proposed in [14] and outlined in Fig. 5.8. It formalises the separation of the general test case description and specification (i.e., *what* needs to be tested and *why*?) from the actual experimental implementation accordingly (i.e., *how* will the system or component test be carried out?).

The test case description (i.e., step ①) is the most abstract and contains predominantly the following attributes:

- The *test objective*, the purpose for carrying out the test or sequence of tests;
- The *Use Case*, which is a high-level description of the functionality of the considered cyber-physical system [15]; and
- the *generic system configuration*, which hierarchically specifies the type of and relation between components and domains.

Then the system under test delineating the system boundaries of the generic system configuration is to be formulated. It encompasses all relevant interactions and function that require investigation. Central in these interactions is the object under investigation, which is the component or sub-system inside the system under test to

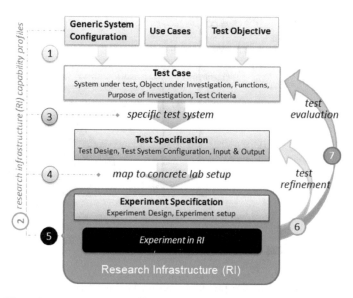

Fig. 5.8 The main specification steps for the holistic testing approach in the ERIGrid research project

which the eventual test criteria shall be evaluated. Test criteria can have a characterisation, verification, or validation nature.

The next step is to project these metrics on a specific test system (i.e., step ③), yielding a description of inputs and outputs, quantitative test metrics to make a test pass or fail, and all further design specifications of the test that are independent of the implementation of the experiment. Some research infrastructures are more tailored for a particular experiment than others. A potentially destructive experiment can only be realised under well-evaluated conditions in specialised laboratories. (Co-)simulation experiments tremendously sustain the qualification of such test boundaries and assumptions. It is hence significant to carefully map a (part of a) particular test on the capabilities of laboratories and institutes (i.e., steps ② and ④). Eventually, the experiment itself needs to be specified. At this level (i.e., ⑤), relevant connection diagrams of lab components, data acquisition, safety regulations, data type conversions (so the experiment setup) and the execution, repetition, and treatment variation (so the experiment design) are significant.

So far the approach has been pretty much feed forward: the test case has been specified at various levels of granularity, starting at a conceptual and abstract dimension to a more detailed, implementation driven level. To account for reproducibility of tests and their statistical relevance, the overall test design is evaluated (for instance by screening or sensitivity analysis) and adjusted accordingly [16] (i.e., steps ⑥ and ⑦).

Fig. 5.9 Illustrative fault
ride-through voltage versus
time curve for wind parks

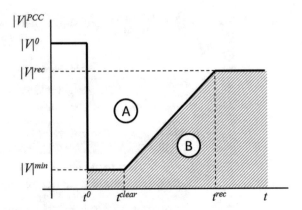

5.3.2   Holistic Test Case Example

The approach outlined in the previous section is illustrated with the test design
that revolves around the use case the so-called fault ride through compliance of an
onshore wind park. As seen from the transmission system, such parks are considered
one single generation entity that must legally comply to various requirements at the
point of common coupling, which is often the high voltage side of the transformer
linking the collection grid with the transmission system. Fault ride-through entails
the ability of the wind power plant to stay connected during voltage dips experienced
at the point of common coupling. This is challenging for various reasons but most
importantly because of the vulnerability of the power electronic converters inside
the individual wind turbines.

The fault ride-through requirement is often expressed by a voltage versus time
envelope as illustrated in Fig. 5.9. Starting at fault ignition at t^0 the wind park is only
permitted to disconnect if the per unit voltage amplitude at the point of common
coupling drops below the indicated voltage profile and enters the dashed gray zone.
Otherwise it should stay connected. It consists of mainly four parts. The prefault part
in which the terminal voltages are around $|V|^0$, the faulted part during which the
(remote) fault causes a severe voltage dip (Ⓐ), the recovery part after fault clearance
(the skew area, Ⓑ), and the post-fault ride through part at which the system and hence
voltage is expected to behave normally again.

In terms of illustrating holistic testing, fault ride through is an attractive option as

- fault ride through involves interactions between multiple domains like ICT,
  physics, control.
- This domain coupling is rigid and these interactions are fast so abstracting away
  phenomena comes at a severe risk of false positives. System Configurations and
  corresponding functions must hence be carefully chosen.
- Though the fault ride through objective is global (at the point of common coupling),
  the implementation is done locally by the individual wind turbines. This needs to
  be reflected into the test metrics (criteria).

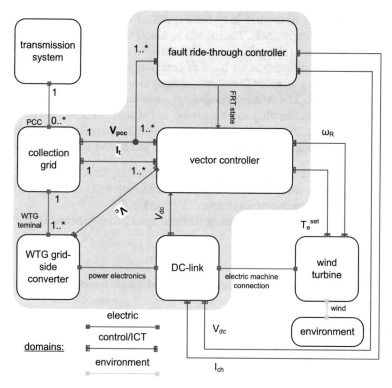

Fig. 5.10 Generic system configuration of the fault ride-through operation of a wind power plant; gray: system under test, yellow: object under investigation

- Testing fault ride through can be destructive to components, which needs a smart test and experiment design.

The objective of the test is to verify the ability of wind park as a generating plant to withstand the external voltage dip and remain connected during and after the fault. The implementation is done locally by the wind turbines, so their controls are the object under investigation. The system functions that are *assumed* dominant to fault ride through are the transient voltage and frequency response of the power system, the active and reactive power controls of the wind turbine, and the power electronics protection devices. The involved (sub-)systems and components of these functions make up the generic system configuration of the test case, which is shown in Fig. 5.10. At this stage only the types and multiplicity of the components and their relational setup is given, all in terms of the functions relevant for fault ride-through. Therefore, this diagram is aimed to be valid for any fault ride-through capable wind park consisting of so-called full-converter generator wind turbines.

The test is considered successful when the wind park is able to stay in operation during the fault and the system quantities like voltage and frequency return dynamically to a stable operating region (i.e., nominal).

Now the generics of the test have been specified we can have a look into the actual design of the experiment. The metrics to assess the test criteria are the output current ($\mathbf{I}_t$) or power to show that the turbines remain connected and the amplitude of the voltage and frequency at the point of common coupling (i.e., $|V|^{PCC}$ and ω_G respectively) to track whether the system returns to a plausible (stable) operating point after the fault. The relevant component input/output variables are shown in Fig. 5.10. As the functionality to be assessed is triggered by short circuits in the transmission system, the design of the test procedure is deterministic and based on the behaviour of the system under test after faults, and can be summarised by:

1. achieve consistent operating point throughout the system under test
2. determine the short circuit location x such that the depth of the voltage dip at the point of common coupling approaches $|V|^{ret}$ in Fig. 5.9.
3. initiate fault at t^0 and start obtaining time-stamped component and system measurements
4. clear the fault at $t^{clear} = y$
5. assess the test criteria
6. vary x and y to cover both short close faults and longer remote faults
7. repeat the experiment.

As field tests are infeasible here for obvious reasons, we need to look for alternative ways to implement the experiment. Ideally speaking, the above test design shall be implemented in a laboratory that at least implements components or subsystems of the wind park in hardware. This could lead to controller hardware-in-the-loop in which for instance the real park controller is assessed whereas the remainder of the system is simulated in real time. Both are then coupled through the I/O interface of the respective real-time simulator. Alternatively, the electrical part of one of the wind turbines could be interfaced with a real time simulator by coupling it through a controllable grid interface (i.e., power hardware-in-the-loop).

Computer simulation is, however, the most optimal option to conduct the experiment. Optimal in the sense of costs, safety (no experiments harmful to humans or hardware) but also flexibility (parameter and model adjustment, determination of initial operating point, model validation, and reproduction). The downside of simulations is that, especially with a rigid cross-domain coupling, the validity of the system response as a whole is determined by the validy of the individual models. To gain a system model with a well defined level of detail for each specialised domain is challenging and usually scale badly (model size, simulation time).

As discussed in previous sections, these issues can largely be overcome with co-simulations, which allow the corresponding subsystems and components to be considered individually by specialised simulators. A master process then keeps track of the interfaces between models over time and event handling. This co-simulation approach is also used with the current example of the wind park.

Figure 5.11 shows a layout of the experimental implementation by co-simulation. It depicts the master algorithm on top and the coupled simutors with their respective subsystems below. The co-simulation employs the functional mockup (FMI) interface

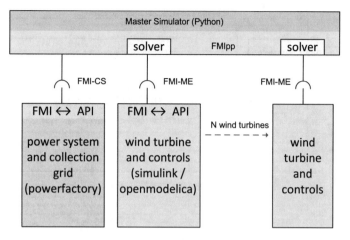

Fig. 5.11 Experimental setup of the co-simulation for fault ride-through testing. FMI: functional mockup interface, CS: co-simulation, ME: model exchange, API: application program interface

standard for interfacing continuous and discrete simulations [17]. FMI defines a set of functions, attributes, and a specification format to which a simulation (FMI for co-simulation) or model (FMI for model exchange) should comply to in order to be able to cooperate/interface with other simulations or the master. A simulator or model that is encapsulated such that it can be programatically linked to the master simulator is called a functional mockup unit. The simulator blocks in Fig. 5.11 each contain an adapter that fulfils this and are hence functional mockup units. The master algorithm is implemented in the Python programming language and uses the Python bindings of the FMI++ library [18], which highly facilitates the adaptation of the functional mockup standard for power system studies.

In this case it was decided to model the power system in Powerfactory whereas the component models are modelled either in Matlab/Simulink or in Openmodelica. Several reasons can be found to have a split like this. First, wind turbine models are commonly vendor-specific and are commonly delivered to customers as black or gray boxed models for a particular simulator. Second, a tool like Powerfactory is well known for its hierarchical scenario and case variation possibilities but is less respected for its ease of model development, which is on its turn the unique feature that makes Simulink and Modelica popular. Third, the static generator model inside Powerfactory allows a flexible and powerful interface to dedicated models (either in its own DSL language or externally developed models).

Although one could in fact represent each wind turbine as an functional mockup unit (shown in Fig. 5.11) and in such a way run a very detailed co-simulation, we limit ourselves to an aggregated wind turbine for illustration purposes. The wind power plant is modelled as a wind turbine aggregate and the collection grid is abstracted to a series impedance, which represents the step-up transformer to the transmission network.

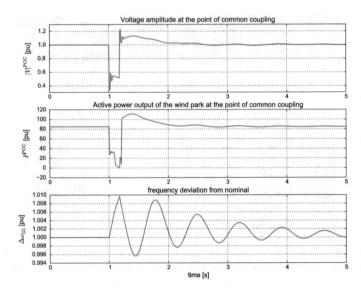

Fig. 5.12 FMI-based co-simulation of a transmission system and full converter based wind turbines; time domain response of a wind park riding through a remote short circuit

The bookkeeping of the interface variables is also accounted for by the master algorithm. Each synchronisation step the grid simulator needs the d–q projection of the static generator reference currents as an input from the wind turbine models, whereas the wind turbine models need the voltage at the point of common coupling ($|V|^{\mathrm{PCC}}$) as an input from the grid simulator.

For one of the tests from outlined the design (nearby fault, interruption after 180 ms) the time domain simulation results are shown in Fig. 5.12. It can be seen that during the fault, the power output of the wind park blocks and the frequency starts to increase as a consequence. After fault clearance the voltages, frequency, and power oscillate and eventually restore to their nominal values. The wind power plant remains connected during the entire simulation yielding a positive test result. During the fault it can be noticed that the voltage amplitude exhibits spikes immediately after fault ignition and clearance, which is probably of numerical nature and caused by the co-simulation experimental setup. This can be resolved by adjusting the synchronisation intervals and/or interface variable handling, and corresponds to step ⑥ in Fig. 5.8. This also holds for testing assumptions made earlier about the aggregation of the wind turbine model at plant level. The outlined holistic testing approach very flexibly enables amendments of the tests and experiments while conserving specifications and definitions at a higher (conceptual) layer.

References

1. P. Palensky, A.A. van der Meer, C.D. López, A. Joseph, K. Pan, Cosimulation of intelligent power systems: fundamentals, software architecture, numerics, and coupling. IEEE Indust. Electron. Mag. **11**(1) (2017)
2. P. Palensky, A.A. van der Meer, C.D. López, A. Joseph, K. Pan, Applied cosimulation of intelligent power systems: implementing hybrid simulators for complex power systems. IEEE Indust. Electron. Mag. **11**(1) (2017)
3. S. Scherfke, S. SchÃtte, *Mosaik-Architecture Whitepaper* (2012), https://mosaik.offis.de/publications/
4. C.D. López, A.A. van der Meer, M. Cvetković, P. Palensky, A variable-rate co-simulation environment for the dynamic analysis of multi-area power systems, in *2017 IEEE Manchester PowerTech* (2017), pp. 1–6
5. ENTSO-E, Common Grid Model Exchange Specification (CGMES), Version 2.5, Draft IEC 61970-600 Part 1, 2nd edn. (2016)
6. ENTSO-E Operational Data Quality Taskforce. Quality of Datasets and Calculations for System Operations, 3rd edn. (2015). Accessed https://docstore.entsoe.eu/ June 2018
7. H. Krishnappa, *Model Validation and Feasibility Analysis of Modelica based Dynamic Simulations using OpenIPSL and CGMES MSc* (Delft University of Technology, Thesis, 2017)
8. K.J. Aström, H. Elmqvist, S.E. Mattsson, Evolution of continuous-time modeling and simulation, in *The 12th European Simulation Multiconference, ESM'98* (Manchester, UK, 1998), pp. 16–19
9. https://fmi-standard.org/
10. M. Cvetković, H. Krishnappa, C.D. López, R. Bhandia, J. Rueda Torres, P. Palensky, Co-simulation and dynamic model exchange with consideration for wind projects, in *Berlin Wind Integration Workshop* (2017), pp. 1–6
11. ERIGrid Project, https://erigrid.eu/
12. J. Trefke, S. Rohjans, M. Uslar, S. Lehnhoff, L. Nordstrom, A. Saleem, Smart grid architecture model use case management in a large European smart grid project, in *Innovative Smart Grid Technologies Europe (ISGT EUROPE)* (Copenhagen, Denmark, 2013), pp. 6–9
13. A.A. van der Meer, P. Palensky, K. Heussen, D.E. Morales Bondy, O. Gehrke, C. Steinbrink, M. Blank, S. Lehnhoff, E. Widl, C. Moyo, T.I. Strasser, V.H. Nguyen, N. Akroud, M.H. Syed, A. Emhemed, S. Rohjans, R. Brandl, A.M. Rohjans, Cyber-physical energy systems modeling, test specification, and co-simulation based testing, in *Workshop on Modeling and Simulation of Cyber-Physical Energy Systems* (Pittsburgh, 2017)
14. M. Blank, S. Lehnhoff, K. Heussen, D.E.M. Bondy, C. Moyo, T. Strasser, Towards a foundation for holistic power system validation and testing, in *2016 IEEE 21st International Conference on Emerging Technologies and Factory Automation (ETFA)* (2016), pp. 1–4
15. International Electrotechnical Comission, *IEC 62559-2:2015 - Use Case Methodology*, https://webstore.iec.ch/publication/22349
16. A.A. van der Meer, C. Steinbrink, K. Heussen, D.E. Morales Bondy, M.Z. Degefa, F. Pröstl, T. Strasser, S. Lehnhoff, P. Palensky, Design of experiments aided holistic testing of cyber-physical energy systems, in *Proceedings of the Modeling and Simulation of Cyber-Physical Energy Systems* (Porto, Portugal, 2018)
17. Modelisar, Functional mock-up interface for co-simulation, MODELISAR (ITEA 2 - 07006), Technical report (2010)
18. The FMI++ Library, http://sourceforge.net/projects/fmipp/, Accessed Feb 2018

Part IV
Coordination and Management of Storage and Flexible Consumption

Chapter 6
Fast Convergence in Electric Vehicle Smart Charging

Sergio Grammatico

Abstract We address the problem to control the charging schedules in a large population of plug-in electric vehicles, considered as heterogeneous noncooperative agents, with different strongly convex cost functions weakly coupled by a common pricing signal, convex charging constraints, e.g. plug-in times, deadlines and capacity limits. We assume a minimal information structure through which a central control unit can broadcast incentive signals to coordinate the decentralized optimal responses of the agents. We propose a dynamic control that, based on fixed-point operator theory, ensures global exponential convergence to an aggregative equilibrium, independently on the population size. We illustrate the benefits of the proposed control via numerical simulations, in scenarios where the aggregate charging demand tends to fill the overnight demand valley. Finally, we touch upon the a generalized setup with convex, separable, joint constraints, e.g. transmission line constraints and shared network resources.

6.1 Introduction

The increasing adoption of plug-in (hybrid) electric vehicles (PEVs) has motivated a wide range of research studies on the impact of high penetration of PEVs in the electrical distribution grid. The motivation is that it is currently envisioned that PEVs will achieve a significant market penetration within the next two decades and uncoordinated charging of large fleets may induce localized overloading, power losses and voltage degradation [1]. To start with, several studies considered centralized optimization methods for scheduling the charging power of PEVs, see e.g. [2, 3] and the references therein. For instance, it is generally convenient to schedule the

This work was partially supported by the Netherlands Organization for Scientific Research (NWO) under research projects OMEGA (grant n. 613.001.702) and P2P-TALES (grant n. 647.003.003).

S. Grammatico (✉)
Faculty of Mechanical, Delft Center for Systems and Control (DCSC), Maritime and Materials Engineering (3mE), Delft University of Technology, Delft, The Netherlands
email: s.grammatico@tudelft.nl
URL: https://sites.google.com/site/grammaticosergio

PEV charging to periods when the overall electricity demand is low, e.g. during the so-called overnight power-demand valley [4, 5].

Since the individual PEV users are foreseen to desire full autonomy, and efficiently scheduling the charging sessions of whole fleets of PEVs has high computational complexity, a plethora of semi-decentralized charging coordination methods have been recently developed [4–11]. Such coordination methods can be regarded as smart charging algorithms indeed. All these methods, each provably correct under specific technical assumptions, consider the PEV users as players, or agents, with individual interests, e.g. in terms of minimal operational cost, including the cost of purchasing the electricity, battery degradation cost, and penalties on the deviation of delivered charge from the desired value, subject to their individual charging needs that can be modeled as hard constraints.

Following up on these setups, we consider a unifying framework and aim at designing an incentive (e.g. pricing) mechanism, namely, a dynamic control law, that is capable to steer the charging/discharging schedules of the PEV agents towards a convenient equilibrium with *linear convergence rate*. As anticipated, the main technical challenge is that the agents are noncooperative, that is, self-interested in minimizing their own cost function, or equivalently, in maximizing their own utility function, and fully heterogeneous in terms of both utility function and charging constraints. Fortunately, since the optimal charging schedule of each PEV agent is postulated to be affected by an incentive signal, which in turn depends on the aggregate charging pattern among all the PEV agents, the theory of *aggregative games* [12–15] is key to analyze the strategic, dynamic interactions among the agents, e.g. in terms of convergence towards a noncooperative game equilibrium, even for large population sizes.

As originally proposed in [4], we assume that the PEV agents determine their optimal charging strategy with respect to an incentive signal (e.g. the electricity price forecast), and the average among these resulting strategies is used to estimate the total demand over the charging horizon. In turn, an updated incentive signal is computed based on such average PEV demand, broadcast to the whole PEV population, and then the process is synchronously repeated in discrete time.

Building upon mathematical tools from variational and convex analysis [16, 17], as well as fixed-point operator theory [18], we refine one of the dynamic control laws proposed in [19–21] to ensure linear convergence rate. We refer to [22] for a continuous-time version of semi-decentralized control for agents playing aggregative games.

Our main technical contribution is to establish global *exponential* convergence of the optimal responses of the controlled agents to an aggregative equilibrium, independently on the population size, on the charging constraints of the PEV agents, and without imposing contraction mapping properties. By considering heterogeneous quadratic cost functions, we generalize the results in [5, 19], and by considering general convex constraints for the charging preferences of the PEV agents, we generalize that in [4, 8, 9] where the constraints are assumed to have a specific structure; in addition, we do not impose any technical assumption on the problem data, e.g. the parameters defining the quadratic cost functions.

The chapter is organized as follows. Sect. 6.2 defines the charging control problem as an aggregative game. Sect. 6.3 presents a dynamic control law with global exponential convergence guarantee. Sect. 6.4 illustrates the performance of the proposed charging control method via numerical simulations. Sect. 6.5 concludes the paper and points at further research directions. Proofs are given in the Appendix.

Basic Notation

$\mathbb{R}$, $\mathbb{R}_{>0}$, $\mathbb{R}_{\geq 0}$ respectively denote the set of real, positive real, non-negative real numbers; $\mathbb{N}$ denotes the set of natural numbers; for $a, b \in \mathbb{N}$, $a \leq b$, $\mathbb{N}[a, b] := [a, b] \cap \mathbb{N}$. $A^{\top} \in \mathbb{R}^{m \times n}$ denotes the transpose of $A \in \mathbb{R}^{n \times m}$. Given vectors $x_1, \ldots, x_T \in \mathbb{R}^n$, $[x_1; \ldots; x_T] \in \mathbb{R}^{nT}$ denotes $\left[x_1^{\top}, \cdots, x_T^{\top} \right]^{\top} \in \mathbb{R}^{nT}$. Given matrices $A_1, \ldots, A_M$, $\mathrm{diag}\,(A_1, \ldots, A_M)$ denotes the block diagonal matrix with $A_1, \ldots, A_M$ in block diagonal positions. With $\mathbb{S}^n$ we denote the set of symmetric $n \times n$ matrices; for a given $Q \in \mathbb{S}^n$, the notations $Q \succ 0$ ($Q \succcurlyeq 0$) and $Q \in \mathbb{S}^n_{\succ 0}$ ($Q \in \mathbb{S}^n_{\succcurlyeq 0}$) denote that Q is symmetric and has positive (non-negative) eigenvalues. I denotes the identity matrix; $\mathbf{0}\,(\mathbf{1})$ denotes a matrix/vector with all elements equal to 0 (1); to improve clarity, we may add the dimension of these matrices/vectors as subscript. $A \otimes B$ denotes the Kronecker product between matrices A and B. Every mentioned set $\mathcal{S} \subseteq \mathbb{R}^n$ is meant to be nonempty. Given $\mathcal{S} \subseteq \mathbb{R}^n$, $A \in \mathbb{R}^{n \times n}$ and $b \in \mathbb{R}^n$, $A\mathcal{S} + b$ denotes the set $\{Ax + b \in \mathbb{R}^n \mid x \in \mathcal{S}\}$. The notation $\varepsilon_N = O(1/N)$ denotes that there exists $c > 0$ such that $\lim_{N \to \infty} N\,\varepsilon_N = c$.

Operator Theory Notation

We denote by $\mathcal{H}_Q$, with $Q \in \mathbb{S}^n_{\succ 0}$, the Hilbert space $\mathbb{R}^n$ with inner product $\langle \cdot, \cdot \rangle_Q :$ $\mathbb{R}^n \times \mathbb{R}^n \to \mathbb{R}$ defined as $\langle x, y \rangle_Q := x^{\top} Q y$ and induced norm $\|\cdot\|_Q : \mathbb{R}^n \to \mathbb{R}_{\geq 0}$ defined as $\|x\|_Q := \sqrt{x^{\top} Q x}$, for all $x, y \in \mathbb{R}^n$. A mapping $f : \mathbb{R}^n \to \mathbb{R}^n$ is Lipschitz continuous relative to $\mathcal{H}_Q$ if there exists $\ell > 0$ such that $\|f(x) - f(y)\|_Q \leq \ell \|x - y\|_Q$ for all $x, y \in \mathbb{R}^n$. f is a contraction (non-expansive) mapping in $\mathcal{H}_Q$ if it is Lipschitz relative to $\mathcal{H}_Q$ with constant $\ell \in (0, 1)$ ($\ell \in (0, 1]$). A mapping $f : \mathbb{R}^n \to \mathbb{R}^n$ is (strictly) monotone in $\mathcal{H}_Q$ if $(f(x) - f(y))^{\top} Q (x - y) \geq 0 \,(> 0)$ for all $x, y \in \mathbb{R}^n$; it is strongly monotone with constant $\ell > 0$ in $\mathcal{H}_Q$ if $(f(x) - f(y))^{\top} Q (x - y) \geq \ell \|x - y\|_Q^2$ for all $x, y \in \mathbb{R}^n$. $\mathrm{Id} : \mathbb{R}^n \to \mathbb{R}^n$ denotes the identity operator, $\mathrm{Id}(x) := x$ for all $x \in \mathbb{R}^n$. A mapping $f : \mathbb{R}^n \to \mathbb{R}^n$ is strongly pseudo-contractive in $\mathcal{H}_Q$ if and only if $\mathrm{Id} - f$ is strongly monotone in $\mathcal{H}_Q$. We implicitly refer to the Hilbert space $\mathcal{H}_I$ unless explicitly mentioned. Given a closed set $C \subseteq \mathbb{R}^n$, the projection operator in $\mathcal{H}_Q$, $\mathrm{proj}^Q_C : \mathbb{R}^n \to C \subseteq \mathbb{R}^n$, is defined as $\mathrm{proj}^Q_C(x) := \arg\min_{y \in C} \|x - y\|_Q^2$ for all $x \in \mathbb{R}^n$.

6.2 Semi-decentralized Smart Charging for Fleets of Plug-in Electric Vehicles

6.2.1 Problem Statement

We consider the charging coordination problem for a large population of $N \gg 1$ noncooperative PEV agents over a time horizon made of multiple charging intervals $\{1, 2, \ldots, n\}$.

We assume that each PEV agent $i \in \mathbb{N}[1, N]$ decides on its charging strategy $u^i = [u^i_1, \ldots, u^i_n]^\top \in \mathcal{U}^i \subset \mathbb{R}^n$, where the set $\mathcal{U}^i$ models its charging constraints.

For instance [4, 5, 8, Equations 1, 2], given the parameters of the state-of-charge dynamics, e.g. charging/discharging rate and efficiency, and the initial state of charge, $\mathcal{U}^i$ represents the time-varying preferences, e.g. upper and lower bounds, on the state of charge and the charging input of the PEV agent i.

We further assume that the PEV agents are noncooperative, meaning that each agent i aims at optimizing its local cost function $J^i : \mathbb{R}^n \times \mathbb{R}^n \to \mathbb{R}$ which depends on (as second argument) a pricing signal common to all agents, without exchanging information with other agents. More formally, for all $i \in \mathbb{N}[1, N]$, we define the *optimal response* mapping $u^{i\star} : \mathbb{R}^n \to \mathbb{R}^n$ as

$$u^{i\star}(\rho) := \arg \min_{y \in \mathcal{U}^i} J^i(y, \rho). \tag{6.1}$$

Technically, throughout the paper we assume compactness and convexity [19, Standing Assumption 1] of the individual charging constraints, and strongly convex quadratic cost functions with affine (linear) dependance on the global pricing variable ρ [8, Sections II, III].

Standing Assumption 1 *Compactness, convexity. The sets $\mathcal{U}^1, \ldots, \mathcal{U}^N \subset \mathbb{R}^n$ are compact and convex. There exists a compact set $\mathcal{U} \subset \mathbb{R}^n$ such that $\cup_{i=1}^N \mathcal{U}^i \subseteq \mathcal{U}$ for all $N \geq 1$.* $\qquad\square$

For simplicity, we consider strongly convex cost functions that are quadratic. The more general case with non-quadratic cost functions can be studied with similar analysis tools, see [23, 24].

Standing Assumption 2 *Strongly convex quadratic cost functions. For all $i \in \mathbb{N}[1, N]$, the cost function $J^i : \mathbb{R}^n \times \mathbb{R}^n \to \mathbb{R}$ in (6.1) is defined as*

$$J^i(y, \rho) := \tfrac{1}{2} y^\top Q^i y + c^{i\top} y + \rho^\top y, \tag{6.2}$$

for some $Q^i \in \mathbb{S}^n_{\succ 0}$ such that $\underline{q} I_n \preccurlyeq Q^i \preccurlyeq \overline{q} I_n$, $\underline{q}, \overline{q} \in \mathbb{R}_{>0}$, and $c^i \in \mathbb{R}^n$. $\qquad\square$

Remark 6.1 The cost function $J^i(u^i, \rho)$ from (6.2) can be seen as the sum of two terms, namely the local penalty term $f^i(u^i) := \frac{1}{2}u^{i\top}Q^iu^i + c^{i\top}u^i$ and the cost of electricity $\rho^\top u^i$. For instance, the local penalty term f^i can model a convex quadratic battery degradation cost as in [8, Equation 8], [9, Equation 5], possibly plus a quadratic penalty $\|u^i - u^i_{\mathrm{ref}}\|^2_{P^i}$ on the deviation from a preferred charging strategy $u^i_{\mathrm{ref}} \in \mathbb{R}^n$, for some weighting matrix $P^i \succcurlyeq 0$.

As an example for this latter addend, the term in [9, Equation 7] reflects the relative importance of delivering the full charge to the PEV i over the charging horizon. Mathematically, that term reads in our notation as $\delta^i\left(\mathbf{1}_n^\top u^i - \gamma^i\right)^2$ for some $\delta^i, \gamma^i > 0$, that is, $\delta^i(u^{i\top}\mathbf{1}_n\mathbf{1}_n^\top u^i - 2\gamma^i\mathbf{1}_n^\top u^i + \gamma^{i2}) = \|u^i - u^i_{\mathrm{ref}}\|^2_{\delta^i\mathbf{1}_n\mathbf{1}_n^\top}$, for some u^i_{ref} such that $\mathbf{1}_n^\top u^i_{\mathrm{ref}} = \gamma^i$.   □

Due to the structure of the quadratic cost function J^i in (6.2), the optimal response of the agents to given incentive signals satisfy the following properties.

Lemma 6.1 *For all $i \in \mathbb{N}[1, N]$, the optimal response mapping $u^{i\star}$ in (6.1) is Lipschitz continuous with constant $\ell := \bar{q}/\underline{q}$ relative to $\mathcal{H}_{I_n}$ and, for all $\rho \in \mathbb{R}^n$, reads as*

$$u^{i\star}(\rho) = \mathrm{proj}^{Q^i}_{\mathcal{U}^i}\left(-Q^{i-1}(\rho + c^i)\right). \tag{6.3}$$

□

In the following, we consider the problem to control via pricing signals ρ the optimal charging strategies $\left(u^{i\star}(\rho)\right)_{i=1}^N$ of the PEV agents towards a set of strategies $(\bar{u}^i)_{i=1}^N$ that satisfy the affine pricing model in [8, Equation 10]:

$$\rho = a\left(d + \tfrac{1}{N}\sum_{i=1}^N \bar{u}^i\right) + b,$$

where $a > 0$ represents the inverse of the price elasticity of demand [19, Section V.C], the vector $d \in \mathbb{R}^n$ represents the normalized non-PEV demand [4, Section III.A, Equation 6], and $b \in \mathbb{R}^n$ represents a constant price addend.

With this aim, let us define the aggregation mapping $\mathcal{A} : \mathbb{R}^n \to \mathbb{R}^n$ as

$$\mathcal{A}(\cdot) := \tfrac{1}{N}\sum_{i=1}^N u^{i\star}(\cdot) \tag{6.4}$$

and pricing mapping $\mathcal{P} : \mathbb{R}^n \to \mathbb{R}^n$ as

$$\mathcal{P}(\cdot) := a(d + \cdot) + b, \tag{6.5}$$

for some given $a \in \mathbb{R}_{>0}$, $d, b \in \mathbb{R}^n$.

The decentralized charging control problem then reads as the design of a pricing vector $\bar{\rho} \in \mathbb{R}^n$ that is a fixed point of the composed mapping $\mathcal{P} \circ \mathcal{A}$ from (6.4), (6.5), i.e., $\bar{\rho} = \mathcal{P}(\mathcal{A}(\bar{\rho}))$.

Existence and uniqueness of such a pricing vector follow by fixed-point theory arguments [25] and are formalized in the following statement.

Proposition 6.1 *Existence and uniqueness of the fixed point. There exists a unique fixed point* $\bar{\rho} \in \mathbb{R}^n$ *of* $\mathcal{P} \circ \mathcal{A}$ *from (6.4), (6.5), i.e.,* $\bar{\rho} = \mathcal{P}(\mathcal{A}(\bar{\rho}))$. $\qquad\qquad\square$

6.2.2 On the Relation of the Fixed Point with Noncooperative and Cooperative Equilibria for Large Population Sizes

The quest for the fixed point of the mapping $\mathcal{P} \circ \mathcal{A}$ has been considered and motivated in the literature [4, 5, 8, 9, 19], as it closely relates to both noncooperative and cooperative equilibria for large population sizes.

Specifically, it follows from [20, Theorem 1] that a fixed point of the mapping $\mathcal{P} \circ \mathcal{A}$ from (6.4), (6.5), i.e., $\bar{\rho} = \mathcal{P}(\mathcal{A}(\bar{\rho}))$, generates a set of strategies $\left(\bar{u}^i := u^{i\star}(\bar{\rho})\right)_{i=1}^N$, with $u^{i\star}$ as in (6.1) for all $i \in \mathbb{N}[1, N]$, that is an aggregative ε_N-Nash equilibrium, where $\varepsilon_N = O(1/N)$, i.e., such that for all $i \in \mathbb{N}[1, N]$, [20, Definition 1]

$$J^i\left(\bar{u}^i, \tfrac{1}{N}\textstyle\sum_{j=1}^N \bar{u}^j\right) \leq \inf_{y \in \mathcal{U}^i} J^i\left(y, \tfrac{1}{N}\left(y + \textstyle\sum_{j \neq i}^N \bar{u}^j\right)\right) + \varepsilon_N.$$

In addition, the existence of an aggregative ($\varepsilon = 0$) Nash equilibrium can be shown as well [24].

On the other hand, it follows from [4, Theorem 6.1], [8, Theorem 3.2], and [26, Propositions 1–3], that the set of optimal responses to a fixed point of the mapping $\mathcal{P} \circ \mathcal{A}$ from (6.4), (6.5), approximates the solution to the social/welfare optimization problem

$$\min_{\boldsymbol{u} \in \mathbb{R}^{nN}} \textstyle\sum_{i=1}^N J^i\left(u^i, \tfrac{1}{N}\textstyle\sum_{j=1}^N u^j\right)$$
$$\text{s.t.} \quad u^i \in \mathcal{U}^i \ \forall i \in \mathbb{N}[1, N]$$

under certain technical assumptions on the cost functions and the constraint sets, for large population size N.

6.3 Exponential Convergence of Smart Charging Control

To control the decentralized optimal responses in (6.1) of the noncooperative agents to a fixed point $\bar{\rho}$ of the mapping $\mathcal{P} \circ \mathcal{A}$ from (6.4), (6.5) through broadcast pricing signals, we conceptually follow the iterative procedure introduced in [4] and refined in [5, 19]. At step $t \in \mathbb{N}$, a central control unit broadcasts a pricing signal $\rho(t) \in \mathbb{R}^n$, to which each PEV agent i responds optimally as $u^{i\star}(\rho(t))$ from (6.1); in turn, based on the average population response $\mathcal{A}(\rho(t))$ from (6.4) and the pricing model $\mathcal{P}(\mathcal{A}(\rho(t)))$ from (6.5), the central control unit computes and broadcast an updated incentive signal $\rho(t+1)$, and then the process is repeated as summarized in Algorithm 3.

For the design of the pricing signal, we propose the dynamic control law

$$\rho(t+1) = \kappa\left(t,\,\rho(t),\,\mathcal{P}\left(\mathcal{A}(\rho(t))\right)\right) \qquad (6.6a)$$
$$= (1-\alpha_t)\,\rho(t) + \alpha_t\,\mathcal{P}\left(\mathcal{A}(\rho(t))\right), \qquad (6.6b)$$

where the positive sequence $(\alpha_t)_{t=0}^{\infty}$, $\alpha_t > 0$ for all $t \in \mathbb{N}$, defines the control law as defined in the next statement.

Definition 6.1 *Dynamic control laws.* The dynamic control law $\kappa : \mathbb{N} \times \mathbb{R}^n \times \mathbb{R}^n \to \mathbb{R}^n$ in (6.6b) is referred as:

Krasnoselskij	$(\kappa \in \mathcal{K})$	if $\alpha_t := \lambda \in (0,1) \; \forall t \geq 0$;
Mann	$(\kappa \in \mathcal{M})$	if $\alpha_t \in \mathbb{R}_{>0}$, $\lim\limits_{t\to\infty} \alpha_t = 0$, $\sum_{t=0}^{\infty} \alpha_t = \infty$.

$\square$

Algorithm 3 Aggregative control of decentralized optimal charging strategies

---

Initialization: $t \leftarrow 0$;

• The controller chooses $\rho_{(0)}$.

Iterate until convergence:

• The controller broadcasts $\rho(t)$ to all the agents.

○ The agents compute in parallel $u^{i\,\star}(\rho(t))$ in (6.1), for all $i \in \mathbb{N}[1, N]$.

• The controller receives $\mathcal{A}(\rho(t))$ in (6.4), and broadcasts

$$\rho(t+1) = \kappa\left(t,\,\rho(t),\,\mathcal{P}\left(\mathcal{A}(\rho(t))\right)\right)$$

as in (6.6b).

$t \leftarrow t + 1$.

Without loss of generality, namely up to a translation of the iteration index, we can assume $\alpha_t \leq 1$ for all $t \in \mathbb{N}$, and indicate $\kappa \in \mathcal{K}$ also if the sequence $(\alpha_t)_{t=0}^{\infty}$ in (6.6b) is such that, for some $\bar{t} \in \mathbb{N}$, $\alpha_t = \lambda \in (0,1)$ for all $t \geq \bar{t}$, e.g. $\alpha_t := \max\{\frac{1}{t+1}, \lambda\}$, for some $\lambda \in (0,1)$. The classes of dynamic control laws $\mathcal{K}$ and $\mathcal{M}$ in Definition 6.1 derive from the Krasnoselskij and Mann fixed point iterations, respectively [18, 19].

Remark 6.2 Consistency of the dynamic control law. A vector $\bar{\rho} \in \mathbb{R}^n$ is a fixed point of the mapping $\mathcal{P} \circ \mathcal{A}$ if and only if, for all $\alpha > 0$, it is a fixed point of the mapping $(1 - \alpha)\mathrm{Id}(\cdot) + \alpha \mathcal{P}(\mathcal{A}(\cdot))$. Therefore, κ in (6.6b) within one of the classes in Definition 6.1 is such that if $\bar{\rho} = \lim_{t \to \infty} \kappa(t, \bar{\rho}, \mathcal{P}(\mathcal{A}(\bar{\rho})))$, then $\bar{\rho} = \mathcal{P}(\mathcal{A}(\bar{\rho}))$, which is a fixed point of $\mathcal{P} \circ \mathcal{A}$. Therefore, the selection of the control law κ in (6.6b) does not affect the set of fixed points of the mapping $\mathcal{P} \circ \mathcal{A}$ in (6.4), which is $\{y \in \mathbb{R}^n \mid y = \mathcal{P}(\mathcal{A}(y))\}$. $\qquad\square$

We note that the static control law $\rho(t + 1) = \mathcal{P}(\mathcal{A}(\rho(t)))$ has been proposed in [4], while the dynamic control law $\kappa \in \mathcal{K}$ in [9], and $\kappa \in \mathcal{M}$ has been proposed in [5]. In all the mentioned setups, it is assumed that the agents' cost functions and constraints are such that the mapping $\mathcal{P} \circ \mathcal{A}$ from (6.4), (6.5) is a contraction mapping.

Our main technical contribution is to prove that for strongly convex cost functions and convex constraints, that is, without assuming nor implying that $\mathcal{P} \circ \mathcal{A}$ is a contraction mapping, nor a nonexpansive mapping, some dynamic control laws $\kappa \in \mathcal{K}$ in (6.6b) ensure that the decentralized optimal responses $\left(u^{i\star}\right)_{i=1}^{N}$ in (6.1) are steered to the desired fixed point *exponentially fast*, that is, with *linear* convergence rate.

Theorem 6.1 (Exponential convergence of Krasnoselskij algorithm) *The dynamic control law $\kappa \in \mathcal{K}$ in (6.6b) with*

$$\alpha_t := \lambda \in \left(0, \frac{1}{2(3 + 3\ell + \ell^2)}\right], \tag{6.7}$$

$\ell := a\bar{q}/\underline{q}$, *ensures that, for all $t \geq \bar{t} \geq 0$ and $\rho_{(0)} \in \mathbb{R}^n$,*

$$\begin{aligned}
\|\rho(t + 1) - \bar{\rho}\| &= \|\rho(t + 1) - \mathcal{P}(\mathcal{A}(\bar{\rho}))\| \\
&= \|\kappa(t, \rho(t), \mathcal{P}(\mathcal{A}(\rho(t)))) - \bar{\rho}\| \\
&\leq \beta^t \|\rho_{(0)} - \bar{\rho}\|,
\end{aligned} \tag{6.8}$$

where $\mathcal{A}$, $\mathcal{P}$ are as in (6.4), (6.5), $u^{i\star}$ as in (6.3) for all $i \in \mathbb{N}[1, N]$, and

$$\beta := 1 - \frac{1}{4(3 + 3\ell + \ell^2)} \in (0, 1). \tag{6.9}$$

Consequently, $\lim_{t \to \infty} \rho(t + 1) = \kappa(t, \rho(t), \mathcal{P}(\mathcal{A}(\rho(t)))) = \bar{\rho} = \mathcal{P}(\mathcal{A}(\bar{\rho}))$. $\qquad\square$

Remark 6.3 The step size λ in (6.7) for the dynamic control law $\kappa \in \mathcal{K}$ depends on the parameters $a, \bar{q}, \underline{q}$ only, not on the population size N, which can be arbitrarily large. Since the control structure is semi-decentralized, the volume of the information exchange and the computational complexity, both at the agent level and at the control unit, are independent on the population size. $\qquad\square$

As corollary to [20, Theorem 2], [21, Theorem 2], we note that the dynamic control law $\kappa \in \mathcal{M}$ in (6.6b) ensures global (not necessarily exponential) convergence to a fixed point of the mapping $\mathcal{P} \circ \mathcal{A}$ from (6.4), (6.5).

Corollary 6.1 (Convergence of Mann algorithm) *The dynamic control law $\kappa \in \mathcal{M}$ in* (6.6b) *ensures that*

$$\lim_{t \to \infty} \rho(t+1) = \kappa\left(t, \rho(t), \mathcal{P}\left(\mathcal{A}(\rho(t))\right)\right) = \bar{\rho} = \mathcal{P}\left(\mathcal{A}(\bar{\rho})\right)$$

for any initial condition $\rho(0) \in \mathbb{R}^n$, where $\mathcal{A}$, $\mathcal{P}$ are as in (6.4), (6.5), *and $u^{i\star}$ as in* (6.3) *for all $i \in \mathbb{N}[1, N]$.* □

Remark 6.4 The dynamic control law $\kappa \in \mathcal{M}$ is fully independent on the problem data, that are the parameters that define the convex quadratic cost functions and the convex charging constraints. Corollary 6.1 shows that the class of model-free dynamic control laws $\kappa \in \mathcal{M}$ ensures global (not necessarily exponential) convergence of the controlled optimal charging strategies towards an aggregative equilibrium, independently on the population size. □

Remark 6.5 It follows from the proofs of Theorem 6.1 and Corollary 6.1 that in our technical setup with strongly convex cost functions with affine pricing and convex constraints, the mapping $\mathcal{P} \circ \mathcal{A}$ belongs to the class of strongly pseudo-contractive mappings [18, Section 4.3], not necessarily to that of contraction mappings, which happens only for a small enough, see also [19, Section V.C]. Therefore, contraction mapping arguments, e.g. via the Banach fixed-point theorem [4, 9], are not applicable to our setup, and in fact persistent oscillations may occur without adopting the appropriate fixed-point iteration in the control unit. □

6.4 Illustrative Numerical Simulations

In this section, we numerically simulate the charging coordination over $n = 20$ time intervals for a population of $N = 10^6$ PEV agents. The nominal values of the numerical parameters defining the cost functions and the charging constraints are taken from [8], and then are randomized as follows to emulate the population variability.

For each PEV agent $i \in \mathbb{N}[1, N]$, we consider the quadratic cost function $J^i(u^i, \rho) = q^i {u^i}^\top u^i + c^i{}^\top u^i + \rho^\top u^i$ that represents the battery degradation cost [8, Section II.C] plus the cost of electricity, where with uniform distribution $q^i \sim \{0.004\} + [-0.001, 0.001]$, $c^i \sim \{0.075\} + [-0.01, 0.01]$. Further, we consider the normalized charging constraints $u^i \in \mathcal{U}^i := [\mathbf{0}_n, \bar{u}^i] \cap \left\{ y \in \mathbb{R}^n \mid \mathbf{1}_n^\top y = \gamma^i \right\}$, where with uniform distribution $\gamma^i \sim \{1\} + [-0.1, 0.1]$, and the vector $\bar{u}^i \in \mathbb{R}^n$ is such that, for all $j \in \mathbb{N}[1, n], \bar{u}^i_j \sim \{0, 0.2\}$, with $\bar{u}^i_j = 0$ (that is, no charging at the time interval

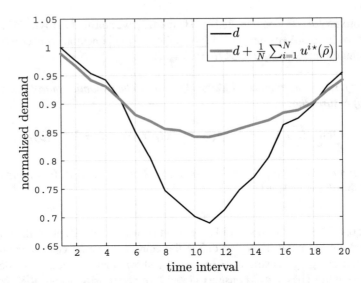

Fig. 6.1 Sum between the normalized background demand d and the average among the charging strategies $\left(u^{i\star}(\bar{\rho})\right)_{i=1}^{N}$ at the equilibrium

j) with probability 20%. In addition, for 10% of the overall population, we consider the V2G option, namely by substituting the lower and upper bounds $[\mathbf{0}_n, \overline{u}^i]$ with $[-\frac{1}{2}\overline{u}^i, \overline{u}^i]$.

We scale the parameters in [8, Section IV] with respect to the population size to derive the price function $\mathcal{P}$ in (6.5), here with parameters $a = 0.038$, $b = 0.061_n$, and vector $d \in \mathbb{R}^n$ being the normalized non-PEV demand, empirically derived from [4, Figure 1], [8, Figure 1].

To apply the result in Theorem 6.1, since $\overline{q}/\underline{q} = 0.005/0.003 = 1.67$, we compute $\ell = a\overline{q}/\underline{q} = 0.063$, thus in (6.7) we have $\lambda_{\max} := \frac{1}{2(3+3\ell+\ell^2)} = 0.1565$.

We numerically compare the convergence performance induced by two control laws $\kappa^\star \in \mathcal{K}$ and $\kappa^\bullet \in \mathcal{M}$, both with global convergence guarantees. For the former, we choose the step sequence $\alpha_t^\star := \max\left\{\frac{1}{t+1}, \lambda_{\max}\right\}$, $t \in \mathbb{N}$, while for the latter we choose the sequence $\alpha_t^\bullet := \frac{1}{t+1}$. We initialize the pricing signal as $\rho_{(0)} := ad + b$.

For the nominal value of the parameters, Fig. 6.1 shows the sum between the normalized background demand and the average among the charging strategies at the equilibrium, i.e., the optimal responses in (6.1) to the fixed point of the mapping $\mathcal{P} \circ \mathcal{A}$. Note that the effect of the V2G operations in the time intervals of high background demand, hence high price of electricity, is that to flatten the overall demand.

Figure. 6.2 in log-log scale supports the theoretical convergence of the sequence $\rho_{(t)}$ that is obtained with the dynamic control laws $\kappa^\star$ (exponential convergence) and $\kappa^\bullet$.

Fig. 6.2 Norm of the difference between the price vector $\rho(t)$ and the equilibrium price $\bar{\rho}$ over the iterations induced by the dynamic control laws $\kappa^\star \in \mathcal{K}$ (solid line) and $\kappa^\bullet \in \mathcal{M}$ (dashed line)

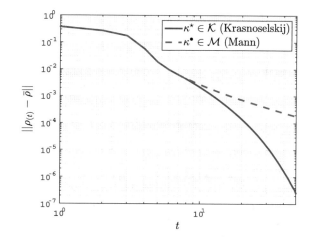

We then set the fixed point convergence criterion to $\|\rho(t) - \mathcal{P}(\mathcal{A}(\rho(t)))\| \leq 10^{-3}$ and run 10^2 simulations, each with parameters randomly taken as described above. In all simulations, we obtain that $\kappa^\star$ induces convergence in 11–13 steps, while $\kappa^\bullet$ in 16–22 steps. Our numerical experience confirms the intuition that $\kappa^\star \in \mathcal{K}$ typically provides faster convergence than control laws in $\mathcal{M}$.

6.5 Conclusion

We have presented a unifying framework for the coordinated charging of noncooperative plug-in electric vehicles in a large population, for which we have proposed a dynamic control law with guaranteed global exponential convergence towards an aggregative equilibrium. The considered framework is applicable far beyond PEV populations, e.g. to general demand side management in the smart grid [27–29].

6.6 Outlook: Semi-decentralized Smart Charging with Joint Constraints

As future research, we shall consider the presence of joint constraints that couple together the optimal decisions of the agents. Joint constraints are indeed motivated by the actual operation with transmission line and network constraints [30–32], and/or shared resources [33].

For simplicity, we can consider affine joint constraints in aggregative form:

$$\frac{1}{N} \sum_{i=1}^{N} u^i \le b. \tag{6.10}$$

Therefore, given a pricing vector ρ, the optimization problem of the agents read as

$$\forall i \in \mathbb{N}[1, N] : \begin{cases} \min_{u^i} J^i(u^i, \rho) \\ \text{s.t. } u^i \in \mathcal{U}^i \\ \quad u^i + \sum_{j \ne i}^{N} u^j \le bN. \end{cases} \tag{6.11}$$

We immediately note that the optimization problems in (6.11) cannot be directly solved with a semi-decentralized information exchange, due to the joint constraint in (6.10). To address this challenge, in the following, we use duality theory. Specifically, for each i, let us define the Lagrangian function $\mathcal{L}^i : \mathbb{R}^n \times \mathbb{R}^n \times \mathbb{R}^n \to \mathbb{R}$ as

$$\mathcal{L}^i(u^i, \rho, \boldsymbol{u}^{-i}, \lambda) := J^i(u^i, \rho) + \lambda^\top \left(-b + \frac{1}{N} \sum_{i=1}^{N} u^i \right)$$
$$= \frac{1}{2} u^{i\top} Q^i u^i + c^{i\top} u^i + \rho^\top u^i + \lambda^\top \left(-b + \frac{1}{N} \sum_{i=1}^{N} u^i \right),$$

where we have imposed that there is only one dual variable λ for all agents, see the discussion on variational equilibria in [23, 34]. Therefore, we have that

$$\nabla_{u^i} \mathcal{L}^i(u^i, \rho, \boldsymbol{u}^{-i}, \lambda) = Q^i u^i + c^i + \rho + \frac{1}{N} \lambda. \tag{6.12}$$

In turn, the *generalized* optimal response mapping of agent i, that is, $\arg\min_{u^i \in \mathcal{U}^i} \mathcal{L}^i$ $(u^i, \rho, \boldsymbol{u}^{-i}, \lambda)$, has the same semi-decentralized structure as in (6.3):

$$u^{i\star}(\rho, \lambda) = \text{proj}_{\mathcal{U}^i}^{Q^i} \left(-Q^i \left(\rho + \frac{1}{N} \lambda + c^i \right) \right). \tag{6.13}$$

In addition, it follows from [23, Proposition 1] that to satisfy the complementarity condition,

$$0 \le \lambda \perp \left(b - \sum_{i=1}^{N} u^{i\star}(\rho, \lambda) \right) \ge 0,$$

the dual variable λ must satisfy the fixed point equation

$$\lambda = \text{proj}_{\mathbb{R}_{\ge 0}^n} \left(\lambda - \gamma \left(b - \sum_{i=1}^{N} u^{i\star}(\rho, \lambda) \right) \right),$$

for some $\gamma > 0$.

The main challenge, which is left as future research avenue, is to design a dynamic control law that combines Krasnoselskij–Mann iterations for the updated of the pricing vector $\rho(t)$, with a projected gradient iteration for the update of $\lambda(t)$, whose role is penalizing the violation of the joint constraint. Alternative semi-decentralized approaches to this generalized aggregative equilibrium have been proposed in [35–37].

Acknowledgements The author would like to thank Mr. Giuseppe Belgioioso (TU Eindhoven) for technical discussions on related topics.

Appendix

Proof of Lemma 6.1

The proof follows from [19, Lemma 4]. First, we have that $-(Q^i)^{-1}(\rho + c^i) = \arg\min_{y \in \mathbb{R}^n} \frac{1}{2} y^\top Q^i y + c^{i^\top} y + \rho^\top y$ as $0 = \frac{\partial}{\partial y}\left(\frac{1}{2} y^\top Q^i y + (\rho + c^i)^\top y\right) = y^\top Q^i + (\rho + c^i)^\top$. Then the following equalities hold:

$$\text{Proj}_{\mathcal{U}^i}^{Q^i}(-(Q^i)^{-1}(\rho + c^i))$$
$$= \arg\min_{y \in \mathcal{U}^i} \left\| y + (Q^i)^{-1}(\rho + c^i) \right\|_{Q^i}^2$$
$$= \arg\min_{y \in \mathcal{U}^i} \frac{1}{2} y^\top Q^i y + c^{i^\top} y + \rho^\top y.$$

The Lipschitz continuity relative to $\mathcal{H}_{I_n}$ with constant $\overline{q}/\underline{q}$ then follows since $\text{Proj}_{\mathcal{U}^i}^{Q^i}$ is Lipschitz continuous relative to $\mathcal{H}_{Q^i}$ with constant 1 [16, Proposition 4.8], hence, for all $x, y \in \mathbb{R}^n$, we have

$$\underline{q}\left\| \text{Proj}_{\mathcal{U}^i}^{Q^i}(-x + c^i) - \text{Proj}_{\mathcal{U}^i}^{Q^i}(-y + c^i) \right\|_{I_n}$$
$$= \left\| \text{Proj}_{\mathcal{U}^i}^{Q^i}(-x + c^i) - \text{Proj}_{\mathcal{U}^i}^{Q^i}(-y + c^i) \right\|_{\underline{q} I_n}$$
$$\leq \left\| \text{Proj}_{\mathcal{U}^i}^{Q^i}(-x + c^i) - \text{Proj}_{\mathcal{U}^i}^{Q^i}(-y + c^i) \right\|_{Q^i}$$
$$\leq \| x - y \|_{Q^i}$$
$$\leq \overline{q}\, \| x - y \|_{I_n},$$

which implies that

$$\left\| \mathrm{Proj}_{\mathcal{U}^i}^{Q^i}(-x+c^i) - \mathrm{Proj}_{\mathcal{U}^i}^{Q^i}(-y+c^i) \right\|_{I_n} \leq \frac{\overline{q}}{\underline{q}} \|x-y\|_{I_n}.$$

∎

Proof of Theorem 6.1

According to Lemma 6.1, the optimal response mapping of agent i reads as in (6.3), i.e., $u^{i\star}(\cdot) = \mathrm{Proj}_{\mathcal{U}^i}^{Q^i}\left(-(Q^i)^{-1}(\cdot+c^i)\right)$.

Since the mapping $\mathrm{Proj}_{\mathcal{U}^i}^{Q^i}$ is monotone in $\mathcal{H}_{Q^i}$ [17, Corollary 12.20] for any closed convex set $\mathcal{U}^i \subseteq \mathbb{R}^n$, we have that, for all $x, y \in \mathbb{R}^m$:

$$\left(\mathrm{Proj}_{\mathcal{U}^i}^{Q^i}\left(-(Q^i)^{-1}(x+c^i)\right) - \mathrm{Proj}_{\mathcal{U}^i}^{Q^i}\left(-(Q^i)^{-1}(y+c^i)\right)\right)^{\top}$$
$$\cdot Q^i\left(-(Q^i)^{-1}(x+c^i)+(Q^i)^{-1}(y+c^i)\right) \geq 0,$$

and equivalently that

$$\left(-u^{i\star}(x)+u^{i\star}(y)\right)^{\top}(x-y) \geq 0.$$

Hence, for all $i \in \mathbb{N}[1, N]$, the mapping $-u^{i\star}(\cdot)$ is monotone in $\mathcal{H}_{I_n}$, and the mapping $\left(\mathrm{Id} - u^{i\star}\right)(\cdot)$ is strongly monotone, as the sum of a strongly monotone mapping and a monotone one [19, Lemma 1]. In turn, also $\frac{1}{N}\sum_{i=1}^{N}\left(\mathrm{Id} - u^{i\star}\right)(\cdot)$ is strongly monotone, and the aggregation mapping

$$\mathcal{A} = \mathrm{Id}(\cdot) - \frac{1}{N}\sum_{i=1}^{N}\left(\mathrm{Id} - u^{i\star}\right)(\cdot) = \frac{1}{N}\sum_{i=1}^{N}u^{i\star}(\cdot)$$

in (6.4) is strongly pseudo-contractive in $\mathcal{H}_{I_n}$ [19, Proof of Theorem 3].

We now consider the composed mapping $\mathcal{P} \circ \mathcal{A} = a\left(d + \frac{1}{N}\sum_{i=1}^{N}u^{i\star}(\cdot)\right) + b$. Since $a > 0$, we have that the mapping $-a\frac{1}{N}\sum_{i=1}^{N}u^{i\star}(\cdot)$ is monotone in $\mathcal{H}_{I_n}$, $\mathrm{Id}(\cdot) - a\frac{1}{N}\sum_{i=1}^{N}u^{i\star}(\cdot) = \mathrm{Id} - a\mathcal{A}$ is strongly monotone, hence from the previous part of the proof, $a\mathcal{A}(\cdot) = \mathrm{Id}(\cdot) - \left(\mathrm{Id}(\cdot) - a\frac{1}{N}\sum_{i=1}^{N}u^{i\star}(\cdot)\right)$ is strongly pseudo-contractive in $\mathcal{H}_{I_n}$. Finally, since $\mathrm{Id} - \mathcal{P} \circ \mathcal{A} = \mathrm{Id} - a\mathcal{A} - ad - b$ is strongly monotone, we conclude that the mapping $\mathcal{P} \circ \mathcal{A} = a\mathcal{A} + ad + b$ is strongly pseudo-contractive as well.

In addition, it follows from Lemma 6.1 that $\mathcal{A}$ is Lipschitz continuous relative to $\mathcal{H}_{I_n}$ with constant $\overline{q}/\underline{q}$, and takes values on the compact set $a\left(d + \frac{1}{N}\sum_{i=1}^{N}\mathcal{U}^i\right) + b \subset \mathbb{R}^n$, hence $\mathcal{P} \circ \mathcal{A}$ is Lipschitz continuous with constant $\ell := a\overline{q}/\underline{q}$.

Therefore, by [18, Corollary 4.2] we have that the Krasnoselskij fixed point iteration, i.e., $\kappa \in \mathcal{K}$ with $\alpha_k := \lambda \in (0, \frac{1}{2(3+3\ell+\ell^2)}]$ ensures global exponential convergence to the fixed point $\bar{\rho}$ of $\mathcal{P} \circ \mathcal{A}$, i.e., (6.8) with β as in (6.9). ∎

Proof of Proposition 6.1

It follows from Lemma 6.1 and the proof of Theorem 6.1 that the mapping $\mathcal{P} \circ \mathcal{A}(\cdot) = a\left(d + \frac{1}{N}\sum_{i=1}^{N} u^{i\star}(\cdot)\right) + b$ is Lipschitz continuous and takes values in a compact set. It then follows from the proof of Theorem 6.1 that the mapping $\mathcal{P} \circ \mathcal{A}$ is strongly pseudo-contractive in $\mathcal{H}_{I_n}$, thus it has one unique fixed point [18, Theorem 4.11]. ∎

Proof of Corollary 6.1

It follows from Lemma 6.1 and Theorem 6.1 that the mapping $\mathcal{P} \circ \mathcal{A}$ is Lipschitz continuous, takes values in a compact set, and is strongly pseudo-contractive in $\mathcal{H}_{I_n}$. Therefore, the Mann fixed point iteration, i.e. $\kappa \in \mathcal{M}$, ensures global convergence to the unique fixed point of $\mathcal{P} \circ \mathcal{A}$ [18, Theorem 4.11]. ∎

References

1. M.D. Galus, M. González Váya, T. Krause, G. Andersson, The role of electric vehicles in smart grids. WIREs Energy Environ. **2**, 384–400 (2013)
2. K. Clement-Nyns, E. Haesen, J. Driesen, The impact of charging plug-in hybrid electric vehicles on a residential distribution grid. IEEE Trans. Power Syst. **5**(1), 371–380 (2010)
3. E. Sortomme, M.M. Hindi, S.D.J. MacPherson, S.S. Venkata, Coordinated charging of plug-in hybrid electric vehicles to minimize distribution system losses. IEEE Trans. Smart Grid **2**(1), 198–205 (2011)
4. Z. Ma, D.S. Callaway, I.A. Hiskens, Decentralized charging control of large populations of plug-in electric vehicles. IEEE Trans. Control Syst. Technol. **21**(1), 67–78 (2013)
5. F. Parise, M. Colombino, S. Grammatico, J. Lygeros, Mean field constrained charging policy for large populations of plug-in electric vehicles, in *Proceedings of the IEEE Conference on Decision and Control* (Los Angeles, California, USA, 2014), pp. 5101–5106
6. S. Grammatico, Exponentially convergent decentralized charging control for large populations of plug-in electric vehicles, in *Proceedings of the IEEE Conference on Decision and Control* (Las Vegas, USA, 2016)
7. L. Gan, U. Topcu, S.H. Low, Optimal decentralized protocol for electric vehicle charging. IEEE Trans. Power Syst. **28**(2), 940–951 (2013)
8. Z. Ma, S. Zou, X. Liu, A distributed charging coordination for large-scale plug-in electric vehicles considering battery degradation cost. IEEE Trans. Control Syst. Technol. **23**(5), 2044–2052 (2015)

9. Z. Ma, S. Zou, L. Ran, X. Shi, I. Hiskens, Efficient decentralized coordination of large-scale plug-in electric vehicle charging. Automatica **69**, 35–47 (2016)
10. F. Parise, S. Grammatico, M. Colombino, J. Lygeros, On constrained mean field control for large populations of heterogeneous agents: decentralized convergence to Nash equilibria, in *Proceedings of the IEEE European Control Conference* (Linz, Austria, 2015)
11. Wu Chenye, Hamed Mohsenian-Rad, Jianwei Huang, Vehicle-to-aggregator interaction game. IEEE Trans. Smart Grid **3**(1), 434–442 (2012)
12. P. Dubey, O. Haimanko, A. Zapechelnyuk, Strategic complements and substitutes, and potential games. Games Econ. Behav. **54**, 77–94 (2006)
13. M. Dindos, C. Mezzetti, Better-reply dynamics and global convergence to Nash equilibrium in aggregative games. Games Econ. Behav. **54**, 261–292 (2006)
14. M.K. Jensen, Aggregative games and best-reply potentials. Econ. Theory, Springer **43**, 45–66 (2010)
15. N.S. Kukushkin, Best response dynamics in finite games with additive aggregation. Games Econ. Behav. **48**(1), 10–94 (2004)
16. H.H. Bauschke, P.L. Combettes, *Convex Analysis and Monotone Operator Theory in Hilbert Spaces* (Springer, Berlin, 2010)
17. R.T. Rockafellar, R.J.B. Wets, *Variational Analysis* (Springer, Berlin, 1998)
18. V. Berinde, *Iterative Approximation of Fixed Points* (Springer, Berlin, 2007)
19. S. Grammatico, F. Parise, M. Colombino, J. Lygeros, Decentralized convergence to Nash equilibria in constrained deterministic mean field control. IEEE Trans. Autom. Control **61**(11), 3315–3329 (2016)
20. S. Grammatico, F. Parise, J. Lygeros, Constrained linear quadratic deterministic mean field control: decentralized convergence to Nash equilibria in large populations of heterogeneous agents, in *Proceedings of the IEEE Conference on Decision and Control* (2015), pp. 4412–4417
21. S. Grammatico, Aggregative control of large populations of noncooperative agents, in *Proceedings of the IEEE Conference on Decision and Control* (Las Vegas, USA, 2016)
22. C. De Persis, S. Grammatico, Continuous-time integral dynamics for aggregative game equilibrium seeking, in *Proceedings of the IEEE European Control Conference*, 2018
23. G. Belgioioso, S. Grammatico, Semi-decentralized Nash equilibrium seeking in aggregative games with coupling constraints and non-differentiable cost functions. IEEE Control Syst. Lett. **1**(2), 400–405 (2017)
24. S. Grammatico, Dynamic control of agents playing aggregative games with coupling constraints. IEEE Trans. Autom. Control **62**(9), 4537–4548 (2017)
25. D.R. Smart, *Fixed Point Theorems* (Cambridge University Press Archive, Cambridge, 1974)
26. M. González Vayá, S. Grammatico, G. Andersson, J. Lygeros, On the price of being selfish in large populations of plug-in electric vehicles, in *Proceedings of the IEEE Conference on Decision and Control* (2015), pp. 6542–6547
27. H. Chen, Y. Li, R.H.Y. Louie, B. Vucetic, Autonomous demand side management based on energy consumption scheduling and instantaneous load billing: an aggregative game approach. IEEE Trans. Smart Grid **5**(4), 1744–1754 (2014)
28. R. Deng, G. Xiao, R. Lu, J. Chen, Fast distributed demand response with spatially- and temporally-coupled constraints in smart grid. IEEE Trans. Smart Grid **11**(6), 1597–1606 (2015)
29. K. Ma, G. Hu, C.J. Spanos, Distributed energy consumption control via real-time pricing feedback in smart grid. IEEE Trans. Control Syst. Technol. **22**(5), 1907–1914 (2014)
30. J. de Hoog, T. Alpcan, M. Brazil, D.A. Thomas, I. Mareels, A market mechanism for electric vehicle charging under network constraints. IEEE Trans. Smart Grid **7**(2), 827–836 (2016)
31. A. Ghavami, K. Kar, S. Bhattacharya, A. Gupta, Price-driven charging of plug-in electric vehicles: Nash equilibrium, social optimality and best-response convergence, in *Proceedings of the Conference on Information Sciences and Systems* (2013)
32. W. Tushar, W. Saad, H.V. Poor, D.B. Smith, Economics of electric vehicle charging: a game theoretic approach. IEEE Trans. Smart Grid **3**(4), 1767–1778 (2012)
33. W. Gu, F. Häusler, W. Griggs, E. Crisostomi, R. Shorten, Smart procurement of naturally generated energy (SPONGE) for PHEVs. Int. J. Control **89**(7), 1467–1480 (2016)

34. F. Facchinei, C. Kanzow, Generalized Nash equilibrium problems. A Q. J. Oper. Res. Springer **5**, 173–210 (2007)
35. G. Belgioioso, S. Grammatico, Projected-gradient algorithms for generalized equilibrium seeking in aggregative games are preconditioned forward-backward methods, in *Proceedings of the IEEE European Control Conference* (2018)
36. S. Grammatico, Aggregative control of competitive agents with coupled quadratic costs and shared constraints, in *Proceedings of the IEEE Conference on Decision and Control* (Las Vegas, USA, 2016)
37. S. Grammatico. An incentive mechanism for agents playing competitive aggregative games, in *Procedings of the International Conference on Network Games, Control and Optimization* (Avignon, France, 2016), pp. 113–122

Chapter 7
Distributed Stochastic Thermal Energy Management in Smart Thermal Grids

Vahab Rostampour, Wicak Ananduta and Tamás Keviczky

Abstract This work presents a distributed stochastic energy management framework for a thermal grid with uncertainties in the consumer demand profiles. Using the model predictive control (MPC) paradigm, we formulate a finite-horizon chance-constrained mixed-integer linear optimization problem at each sampling time, which is in general non-convex and hard to solve. We then provide a unified framework to deal with production planning problems for uncertain systems, while providing a-priori probabilistic certificates for the robustness properties of the resulting solutions. Our methodology is based on solving a random convex program to compute the uncertainty bounds using the so-called scenario approach and then, solving a robust mixed-integer optimization problem with the computed randomized uncertainty bounds at each sampling time. Using a tractable approximation of uncertainty bounds, the proposed formulation retains the complexity of the problem without chance constraints. We also present two distributed approaches that are based on the alternating direction method of multipliers (ADMM) to solve the robust mixed-integer problem. The performance of the proposed methodology is illustrated using Monte Carlo simulations and employing two different problem formulations: optimization over input sequences (open-loop MPC) and optimization over affine feedback policies (closed-loop MPC).

This research was supported by the Uncertainty Reduction in Smart Energy Systems (URSES) research program funded by the Dutch organization for scientific research (NWO) and Shell under the Aquifer Thermal Energy Storage Smart Grids (ATES-SG) project with grant number 408-13-030.

V. Rostampour (✉) · T. Keviczky
Delft Center for Systems and Control, Delft University of Technology, Mekelweg 2, 2628 CD Delft, The Netherlands
e-mail: v.rostampour@tudelft.nl

T. Keviczky
e-mail: t.keviczky@tudelft.nl

W. Ananduta
Automatic Control Department, Universitat Politècnica de Catalunya, Institut de Robòtica i Informàtica Industrial (IRI, CSIC-UPC), Barcelona, Spain
e-mail: wananduta@iri.upc.edu

© Springer Nature Switzerland AG 2019
P. Palensky et al. (eds.), *Intelligent Integrated Energy Systems*,
https://doi.org/10.1007/978-3-030-00057-8_7

141

7.1 Introduction

Smart Thermal Grids (STGs) represent a new concept in the energy sector that involves the use of the smart grid concept in thermal energy networks connecting several households and greenhouses (agents) to each other via a transport line of thermal energy. One of the major challenges in sustainable energy systems is to improve the efficiency, reliability, and sustainability of the production and the distribution of energy. STGs can contribute to obtaining sustainable energy systems by introducing a reliable production plan using renewable energy sources such as solar or geothermal energy and provide efficient large-capacity storage options. This results in a reduction of carbon dioxide (CO_2) emissions, improved energy efficiency, and the implementation of renewable energy systems [1].

In an STG setting, the agents have a potential to contribute to the overall energy balance. Every agent fulfills the role of a consumer when it demands more energy than it produces with its production units (e.g. micro-combined heat and power), and fulfills the role of a producer when the demand is less than the production of its production units [2]. Since the major energy consumption is typically used for thermal purposes, the motivation for STGs can be both economical and environmental. A better price is achieved with less energy transport when the resources are used more efficiently, while the thermal energy losses are reduced.

We therefore foresee a shift towards a situation where a large number of small scale agents (e.g. utility companies and independent users) have more impact on the energy balance of the grid, while their optimal decisions are made by considering the thermal demand profiles, which are uncertain. The planning of thermal energy production to match supply and demand is challenging since predictions on the thermal energy demand are not perfect. This highlights the necessity of formulating stochastic variants of standard day-ahead planning problems in the grid, while providing probabilistic guarantees regarding the satisfaction of smart grid system constraints.

Model predictive control (MPC) is one of the most widely used advanced control design methods that can handle constraints on both inputs and states, and can obtain an optimal control sequence that minimizes a given objective function subject to the model and operational constraints in a receding horizon fashion. One way to treat uncertainty is to use a robust MPC formulation [3–5], which provides a control law that satisfies the problem constraints for all admissible uncertain variables by assuming that the uncertainty is bounded. However, the resulting solution tends to be conservative since all uncertainty realizations are treated equally. Stochastic MPC offers an alternative approach to achieve a less conservative solution, thereby the system constraints are treated in a probabilistic sense (chance constraints), meaning that the constraints need to be satisfied only probabilistically up to a pre-assigned level to reduce the conservatism of robust MPC. An effective solution to address such problems is to employ randomized algorithms that require substituting the chance constraint with a finite number of hard constraints corresponding to samples of the uncertainty set. Randomized MPC approximates stochastic MPC via the so-called scenario approach (see [6] and the references therein), and if the underlying optimization problem is convex with respect to the decision variables, finite sample guarantees can be provided for a desired confidence level of constraint fulfillment.

In this work we cannot employ the well-known scenario approach due to the fact that the underlying problem is not convex (mixed-integer program). The main challenge here is in the presence of the uncertain thermal energy demands to compute a discrete (binary) variable vector that corresponds to the on-off status of the generating units, and a continuous variable vector that is related to the amount of thermal energy that each unit should produce to satisfy a given demand level at each sampling time. Instead, we propose a two-step procedure that is based on a mixture of randomized and robust optimization [7]. We first determine a probabilistic bounded set of uncertainties that is guaranteed to include a given percentage of uncertainty realizations. Then, we use the obtained set in a deterministic robust MPC approach. Note that the first step leads to a convex sub-problem even if the original problem contains binary variables. In this way we can have similar results in terms of confidence level of constraint fulfillment as in the standard scenario approach. Using a tractable approximation of uncertainty bounds, the deterministic robust formulation leads to a tractable problem for each sampling time. A framework for stochastic linear systems using a combination of randomization and robust optimization was introduced in [8]. In this work instead we introduce a new framework for stochastic hybrid linear systems that leads to stochastic mixed-integer optimization problems. Due to the large number of agents in a large-scale multi-agent setting as in this work, computational burden may also become an issue. In this line, distributed approaches are seen to be more suitable, and also, more flexible and scalable than the centralized counterpart [9]. We therefore propose two distributed schemes that can be applied to solve the resulting tractable problem using the alternating direction method of multipliers (ADMM). The main contributions of this work are as follows:

- A technical description of smart thermal grids with uncertainties in the consumer demand profiles as an optimization problem formulation.
- The problem formulation leads to a finite-horizon chance-constrained mixed-integer linear program at each sampling time. To solve this problem, we first formulate an auxiliary problem to obtain a bounded set for the uncertainty. Using the scenario approach, the result of this sub-problem is a subset of the uncertainty space that contains a portion of the probability mass of the uncertainty with high confidence level. We then solve a robust version of the initial problem subject to the uncertainty confined in the obtained set. Note that our method does not restrict the underlying probability distribution of uncertainty as in robust optimization methods and it is only assumed that the uncertainties are independent and identically distributed.
- To guarantee that the resulting problem is solvable, we develop a tractable scheme based on the dependency of the constraint functions on the uncertainty sets.
- Both the open-loop stochastic MPC formulation and the closed-loop affine feedback policies of stochastic MPC formulation are described and used to illustrate a performance of the proposed methodology using Monte Carlo simulations.

- We provide two distributed ADMM schemes to solve the resulting tractable optimization problem: (1) a fully distributed and (2) a distributed with coordination schemes.

It is important to highlight that this work is based on the conference paper published in [10] and thesis report in [11]. The layout of this work is as follows: Sect. 7.2 provides a general stochastic MPC framework for the problem of uncertain smart thermal grids. In Sect. 7.3 a tractable methodology is developed and probabilistic performance guarantees are provided. The distributed ADMM schemes are formulated in 7.4. In Sect. 7.5, we demonstrate the efficiency of the proposed methodology through a numerical example. Finally, Sect. 7.6 provides some concluding remarks and directions for future work.

7.2 Problem Formulation

This section provides a brief description of smart thermal grids with multiple agents that can be producers and consumers of power and heat in a smart grid setting. The goal of the agents is to match the local consumption and production to avoid transport losses in the network and improve energy efficiency.

System Description

Consider a regional thermal grid consisting of N agents (households, greenhouses). We describe the model of a single agent that is facilitated with a micro-combined heat and power plant (micro-CHP), a boiler, and a heat storage. Each agent can be both producer and consumer which is known as the prosumer concept. This model introduces the technical constraints of each agent and the coupling between such agents in the network. Moreover, for every transaction of thermal energy in the smart grid, there are several heat exchangers located near the corresponding agents and we assume that the heat exchangers do not add additional costs to the heat production for the sake of simplicity.

For a day-ahead planning production problem of each agent, we consider a finite horizon $N_t = 24$ problem with hourly steps, and introduce the subscript t in our notation to characterize the value of the quantities for a given time instance $t \in \{0, 1, \ldots, N_t - 1\}$. For each sampling time t of the problem for all agents $i \in \{1, 2, \ldots, N\}$, we define the main vector of control decision variables to be

$$u_{i,t}^m := [p_{g,t}, p_{ug,t}, p_{dg,t}, h_{g,t}, h_{b,t}, h_{im,t}, c_{g,t}^{su}, c_{b,t}^{su}, z_{g,t}, z_{b,t}]^\top \in \mathbb{R}^{10},$$

where $p_{g,t}, h_{g,t}$ denote the electrical power and heat production by the micro-CHP, $p_{ug,t}, p_{dg,t}$ relate to the up and down spinning of electrical power by the micro-CHP, $h_{b,t}, h_{im,t}$ correspond to the heat provided by the boiler and imported heat from external parties during period of high heat demand, $c_{i,t}^{su} := [c_{g,t}^{su}, c_{b,t}^{su}]^\top$ is a vector that contains the startup cost of micro-CHP and boiler, and $z_{i,t} := [z_{g,t}, z_{b,t}]^\top$ are auxiliary variables needed to model the minimum up and down times of each micro-CHP and boiler, respectively. Moreover, $v_{i,t} : [v_{g,t}, v_{b,t}]^\top \in \{0, 1\}^{N_v = 2}$ is a binary vector of dimension 2 and denotes the on-off status of each micro-CHP and boiler for each agent i at step t. We call the difference between the level of heat storage and the forecast of heat demand $h^f_{d,t}$ the imbalance error $x_{i,t} \in \mathbb{R}^{N_x = 1}$ at agent i, defined as

$$x_{i,t} = h_{s,t} - h^f_{d,t} , \tag{7.1}$$

where $h_{s,t}$ represents the heat storage level (assuming there are no thermal losses in the conversion and storage system). The heat storage level has the following dynamics:

$$h_{s,t+1} = \eta_s x_{i,t} + \eta_s \left(h_{g,t} + h_{b,t} + h_{im,t} + \sum_{j \in N_{-i}} (1 - \alpha_{ij}) h^{ij}_{ex,t} \right) ,$$

where $\eta_s \in (0, 1)$ and $\alpha_{ij} \in (0, 1)$ denote the efficiency of storage and the heat loss coefficient due to transportation between agent i and j, respectively. N_{-i} is the set of neighbors of agent i and is given by

$$N_{-i} \subseteq \{1, 2, \ldots, N\} \backslash \{i\} .$$

In order for an agent to contribute to the local balancing of heat by exchanging heat with neighbors, we define an auxiliary control variable vector $u^a_{i,t} \in \mathbb{R}^{|N_{-i}|}$ with elements $h^{ij}_{ex,t}$ denoting the exchanged heat between agent i and other adjacent agents $j \in N_{-i}$. Notice that $h^{ij}_{ex,t}$ can have either positive or negative values depending on if agent i imports or exports heat from or to agent j, respectively. By substituting $h_{s,t}$ in (7.1), one can derive the dynamical behavior of imbalance $x_{i,t}$ that is given by

$$x_{i,t+1} = A_i x_{i,t} + B_i u_{i,t} + w_{i,t} , \tag{7.2}$$

where $B_i = \eta_s [b_1^\top, b_2^\top]^\top$, with $b_1 = [0, 0, 0, 1, 1, 1, 0, 0, 0, 0] \in \mathbb{R}^{10}$, $b_2 \in \mathbb{R}^{|N_{-i}|}$ containing elements $(1 - \alpha_{ij})$, and $A_i = \eta_s$. The complete vector of control decision variables is $u_{i,t} = [u^{m\top}_{i,t}, u^{a\top}_{i,t}]^\top \in \mathbb{R}^{N_{u_i}}$ with $N_{u_i} = 10 + |N_{-i}|$ for every agent i at each step t. By definition (7.1), $w_{i,t} := -h^f_{d,t+1} \in \mathbb{R}^{N_w = 1}$ corresponds to the forecast heat demand in the next time step. We now consider that the only uncertainty is due to the deviation of the actual heat demand from its forecast value and therefore, $w_{i,t}$ represents an uncertain parameter for every hour and for each agent.

Our goal is to find the control input $u_{i,t}$ for each agent i such that the imbalance error stays a small positive value for all steps $t \in \{0, 1, \ldots, N_t - 1\}$ at minimal production cost and satisfying physical constraints. We associate an economic linear cost function with each agent i at step t as

$$J_i(u_{i,t}) = c_i^\top u_{i,t}^m ,$$ (7.3)

where c_i is a cost vector and is defined as

$$c_i := [c_{gas}\eta_{CHP}^{-1}, c_{up}, c_{dp}, 0, c_b\eta_b^{-1}, c_{im}, 1, 1, 0, 0]^\top \in \mathbb{R}^{10}.$$

c_{gas} relates to the cost of natural gas that is used by micro-CHP and $\eta_{CHP}, \eta_b \in (0, 1)$ are the efficiency of a micro-CHP and a boiler at each step, respectively. c_{up}, c_{dp} denote the cost of up and down spinning electrical power production by micro-CHP, respectively. We define $p_{ug,t}, p_{dg,t}$ to be up and down spinning variables that are related to the amount of surplus and needed electrical power, respectively, in each agent at each step with respect to the local power demand. The cost of heat generated by a boiler is c_b, and the cost of imported heat from an external party is c_{im}. The seventh and eighth entry of c_i represent the start-up costs. The cost associated with the thermal energy produced by the micro-CHP is considered to be zero due to the fact that the electrical power and thermal energy generated by a micro-CHP are coupled by $h_{g,t} = \frac{\eta_h}{\eta_p} p_{g,t}$ where $\eta_h, \eta_p \in (0, 1)$ are the efficiency of a micro-CHP for production of thermal energy and electrical power, respectively.

The resulting optimization problem for an agent i is given by

$$\min_{\{u_{i,t}, v_{i,t}\}_{t=0}^{N_t-1}} \sum_{t=0}^{N_t-1} J_i(u_{i,t})$$ (7.4a)

subject to:

1. Startup cost constraints for $t = 0, 1, \ldots, N_t - 1$:

$$c_{i,t}^{su} \geq \Lambda^{su}(v_{i,t} - v_{i,t-1}) , \quad c_{i,t}^{su} \geq 0 ,$$ (7.4b)

   where Λ^{su} is a diagonal matrix including the startup costs of each micro-CHP and boiler.

2. Production and transportation capacity constraints for $t = 0, 1, \ldots, N_t - 1$:

$$v_{g,t} p_g^{min} \leq p_{g,t} \leq p_g^{max} v_{g,t} ,$$ (7.4c)

$$v_{g,t} h_g^{min} \leq h_{g,t} \leq h_g^{max} v_{g,t} ,$$ (7.4d)

$$v_{b,t} h_b^{min} \leq h_{b,t} \leq h_b^{max} v_{b,t} ,$$ (7.4e)

$$h_{im,t}^{min} \leq h_{im,t} \leq h_{im,t}^{max} ,$$ (7.4f)

$$h_{ex,t}^{min} \leq h_{ex,t}^{ij} \leq h_{ex,t}^{max} , \quad \forall i, j \in N_{-i}$$ (7.4g)

where p_g^{min}, p_g^{max} denote the minimum and maximum electrical power production capacities of each micro-CHP, h_g^{min}, h_g^{max} relate to the minimum and maximum heat production capacities of each micro-CHP, h_b^{min}, h_b^{max} are the minimum and maximum heat production capacities of each boiler, $h_{im}^{min}, h_{im}^{max}$ are the minimum and maximum available heat capacities of each external party, $h_{\text{ex}}^{\min}$, $h_{\text{ex}}^{\max}$ represent the minimum and maximum transportation capacities of neighbors.

3. Balance constraints for heat exchanged with neighbors for $t = 0, 1, \ldots, N_t - 1$:

$$h_{\text{ex},t}^{ij} + h_{\text{ex},t}^{ji} = 0 \ . \ \forall i, j \in N_{-i} \tag{7.4h}$$

4. Up and down spinning electrical power constraints for $t = 0, 1, \ldots, N_t - 1$:

$$-p_{\text{dg},t} \leq p_{\text{g},t} - p_{\text{d},t} \leq p_{\text{ug},t}, \ p_{\text{ug},t} \geq 0, \ p_{\text{dg},t} \geq 0, \tag{7.4i}$$

where $p_{\text{d},t}$ is a local electrical power demand for each agent $i \in \{1, \ldots, N\}$.

5. Ramping capacity constraint for all $t = 0, 1, \ldots, N_t - 1$:

$$-p_{\text{g}}^{\text{down}} \leq p_{\text{g},t} - p_{\text{g},t-1} \leq p_{\text{g}}^{\text{up}}, \tag{7.4j}$$

where $p_{\text{g}}^{\text{down}}$, p_{g}^{up} denote the down and up capacity of decreasing and increasing electrical power of a micro-CHP within two consecutive periods, respectively. Note that this constraint is considered just for the electrical power of micro-CHP due to the fact that heat can be produced within each step.

6. Status change constraints for $t = 0, 1, \ldots, N_t - 1$:

$$z_{i,t} \geq v_{i,t} - v_{i,t-1} , \ z_{i,t} \geq 0 ,$$

$$\sum_{\tau=t+1-\Delta t_{\text{up}}}^{t} z_{i,\tau} \leq v_{i,t} , \ \forall t \in \{\Delta t_{\text{up}}, \ldots, N_t - 1\} ,$$

$$\sum_{\tau=t+1}^{t+\Delta t_{\text{down}}} z_{i,\tau} \leq 1 - v_{i,t} , \ \forall t \in \{1, \ldots, N_t - 1 - \Delta t_{\text{down}}\} , \tag{7.4k}$$

where Δt_{up}, $\Delta t_{\text{down}} \in \mathbb{R}_+$ denote the minimum time an agent needs to change status of the micro-CHP and the boiler.

7. Probabilistic constraint:

$$\mathbb{P}(x_{i,t+1} \geq 0 , \ \forall t \in \{0, 1, \ldots, N_t - 1\}) \geq 1 - \epsilon , \tag{7.4l}$$

where $\epsilon \in (0, 1)$ is the admissible constraint violation parameter. Given $x_{i,0}$ is equal to $x_i(k)$ which is the current state measurement at real time k. This constraint implies that the imbalance error should be a positive value at minimum production cost for all heat demand realizations with high probability $1 - \epsilon$.

The proposed optimization problem (7.4) is a finite-horizon multi-stage, chance-constrained mixed-integer linear program, whose stages are coupled by the startup binary (7.4b), ramping (7.4j), status change (7.4k) and imbalance error (7.4l) constraints.

Open-Loop Stochastic MPC

In order to formulate a stochastic MPC for the overall smart thermal grid imbalance problem, we first extend the optimization problem (7.4) for all agents in the grid. Let us define $X_t := [x_{1,t}^\top, \ldots, x_{N,t}^\top]^\top \in \mathbb{R}^{N_x N}$, $U_t := [u_{1,t}^\top, \ldots, u_{N,t}^\top]^\top \in \mathbb{R}^{N_u}$, where $N_u = \sum_{i=1}^N N_{u_i}$, and $V_t := [v_{1,t}^\top, \ldots, v_{N,t}^\top]^\top \in \mathbb{R}^{N_v N}$ to be the state, control input and binary variables of the grid, respectively. We also define $W_t := [w_{1,t}^\top, \ldots, w_{N,t}^\top]^\top \in \mathbb{R}^{N_w N}$ to be an uncertainty vector of all the agents. The grid cost $J(U_t)$ at step t is assumed to be the sum of the individual costs for all agents

$$J(U_t) = C^\top U_t = \sum_{i=1}^N J_i(u_{i,t}) \,,$$

where $C := [c_1^\top, \ldots, c_N^\top]^\top$. The dynamics of the imbalance error of all agents in the grid can be expressed as

$$X_{t+1} = AX_t + BU_t + W_t \,, \tag{7.5}$$

where $A = \text{diag}(A_1, \ldots, A_N)$ and $B = \text{diag}(B_1, \ldots, B_N)$. The uncertain variable vector $W_t \in \mathbb{R}^N$ is defined on a probability space Δ. It is assumed that Δ is endowed with the Borel $\sigma-$algebra and $\mathbb{P}$ is a probability measure defined over Δ. It is important to note that for our study we only need a finite number of instances of W_t, and we do not require the probability space Δ and the probability measure $\mathbb{P}$ to be known explicitly.

To illustrate the advantages obtained by adopting the policies that were discussed at the end of the preceding section, we first need to introduce some compact notations for the overall system dynamics evolution along the finite time horizon. Consider the following vectors of state, control input, binary variables, and uncertainty parameter matrices.

$$\mathbf{X} = \begin{bmatrix} X_1 \\ X_2 \\ \vdots \\ X_{N_t} \end{bmatrix}, \ \mathbf{U} = \begin{bmatrix} U_0 \\ U_1 \\ \vdots \\ U_{N_t-1} \end{bmatrix}, \ \mathbf{V} = \begin{bmatrix} V_0 \\ V_1 \\ \vdots \\ V_{N_t-1} \end{bmatrix}, \ \mathbf{W} = \begin{bmatrix} W_0 \\ W_1 \\ \vdots \\ W_{N_t-1} \end{bmatrix}.$$

The imbalance error dynamics for all agents over the prediction horizon can be now written as

$$\mathbf{X} = \mathbf{A}X_0 + \mathbf{B}\mathbf{U} + \mathbf{H}\mathbf{W} \,,$$

where

$$\mathbf{A} = \begin{bmatrix} A \\ A^2 \\ \vdots \\ A^{N_t} \end{bmatrix}, \quad \mathbf{B} = \begin{bmatrix} B & 0 & \cdots & 0 \\ AB & B & \ddots & 0 \\ \vdots & & \ddots & \ddots & 0 \\ A^{N_t-1}B & \cdots & AB & B \end{bmatrix}, \quad \mathbf{H} = \begin{bmatrix} I & 0 & \cdots & 0 \\ A & I & \ddots & 0 \\ \vdots & & \ddots & \ddots & 0 \\ A^{N_t-1} & \cdots & & A & I \end{bmatrix}.$$

The initial state values are defined by $X_0 := X(k) = [x_1^\top(k), \dots, x_N^\top(k)]^\top \in \mathbb{R}^{N_x N}$ and it is assumed that the current state measurement at real time k is given to each agent together with the the forecasted thermal energy demand for the next day step-wise. The objective function can be expressed by $\mathbf{J}(\mathbf{U}) = \mathbf{C}^\top \mathbf{U}$, where $\mathbf{C} = \mathbf{1}^{N_t} \bigotimes C$ using the Kronecker product.

We are now in a position to define the optimization problem for the overall smart thermal grid as follows:

$$\min_{\mathbf{U}} \quad \mathbf{J}(\mathbf{U}) \tag{7.6a}$$

$$\text{s.t.} \quad \mathbb{P}_{\mathbf{W}}(\mathbf{A}X_0 + \mathbf{B}\mathbf{U} + \mathbf{H}\mathbf{W} \geq 0) \geq 1 - \epsilon, \quad X_0 = X(k), \tag{7.6b}$$

$$\mathbf{E}\mathbf{U} + \mathbf{F}\mathbf{V} + \mathbf{P} \leq 0, \quad \mathbf{W} \in \Delta^N \tag{7.6c}$$

where $\mathbf{E}$, $\mathbf{F}$ and $\mathbf{P}$ are matrices of appropriate dimensions. Notice that $\mathbb{P}_{\mathbf{W}}$ depends on the string of uncertain scenario realizations. The solution of (7.6) is the optimal planned input sequence $\{U_0^*, V_0^*, \dots, U_{N_t-1}^*, V_{N_t-1}^*\}$. Based on the MPC paradigm the current input is set to $\{U(k), V(k)\} := \{U_0^*, V_0^*\}$ and we proceed in a receding horizon fashion. This means (7.6) is solved at each step t by using the current measurement of the state $X(k)$. Due to the presence of chance constraints, the feasible set is in general non-convex and hard to determine explicitly. We describe a tractable formulation to solve (7.6) by using robust randomization techniques in Sect. 7.3.

Closed-Loop Stochastic MPC

In the presence of uncertainty, the problem of finding the optimal state feedback policy becomes quite challenging. One way to tackle this problem is to look for a sub-optimal solution by parameterizing the control input variables. As a first approach, one can directly parameterize the control input variables as an affine function of the uncertainty

$$U_t = \overline{U}_t + \sum_{j=0}^{t-1} \theta_{t,j} W_j,$$

where $\overline{U}_t$ and $\theta_{t,j}$ are optimization variables. In this way the designed closed-loop control system is equivalent to an open-loop control system with a feedforward uncertainty compensator [12, 13]. Consider the following finite-horizon chance-constrained mixed-integer linear program by adopting a feedback control policy (7.7d) that is affine in the uncertainty samples.

$$\min_{\overline{U},G,V} \ J(U) \tag{7.7a}$$

$$\text{s.t.} \ \ \mathbb{P}_W(AX_0 + BU + HW \geq 0) \geq 1 - \epsilon \ , \ X_0 = X(k) \ , \tag{7.7b}$$

$$EU + FV + P \leq 0 \ , \tag{7.7c}$$

$$U = \overline{U} + GW \ , \ W \in \Delta^N \tag{7.7d}$$

where matrices $\overline{U}$ and G are given by

$$\overline{U} = \begin{bmatrix} \overline{U}_0 \\ \overline{U}_1 \\ \vdots \\ \overline{U}_{N_t-1} \end{bmatrix} , \ G = \begin{bmatrix} 0 & 0 & \cdots & 0 \\ \theta_{1,0} & 0 & \ddots & 0 \\ \vdots & \ddots & \ddots & 0 \\ \theta_{N_t-1,0} & \cdots & \theta_{N_t-1,N_t-2} & 0 \end{bmatrix} .$$

Notice that each element of G has dimension $\mathbb{R}^{N_u \times N_w N}$.

To solve (7.6) and (7.7), we have to transform the chance-constrained problem to a tractable one without introducing any assumptions on $\mathbb{P}$ and its moments. Hence, we follow a randomization-based approach. The proposed procedure in [6] is called scenario approach that allows to substitute the chance constraints with a finite number of hard constraints corresponding to scenarios of the uncertainty and provides a probabilistic guarantee, if the underlying problem is convex with respect to the decision variables. The number of scenarios of the uncertainty realizations N_s that needs to be extracted must satisfy

$$N_s \geq \frac{2}{\epsilon} \left(d + \ln \frac{1}{\beta} \right) , \tag{7.8}$$

where $\epsilon \in (0, 1)$ is a desired level of constraint violations, d is the number of decision variables and $\beta \in (0, 1)$ is a desired confidence level with which the drawn scenarios lead to a feasible solution [14, 15]. Unfortunately, we cannot follow this approach, due to the binary vector V. Even if the convexity condition were satisfied, the number of scenarios that one needs to generate grows linearly with the number of decision variables, thus hampering the applicability of the method to large-scale problems [8]. For example, the number of decision variables in the proposed open-loop stochastic MPC formulation (7.6) is $d = (N_u + N_v N)N_t$, and the number of decision variables in the proposed closed-loop affine uncertainty feedback policy stochastic MPC formulation (7.7) is $d + d_G$, where $d_G = N_u N_w N \frac{(N_t-1)N_t}{2}$. Due to the high dimension of decision space, we cannot even employ the extensions to non-convex problems in [17]. To overcome this difficulty, we propose a tractable methodology based on the results of [7] in Sect. 7.3.

7.3 Centralized Stochastic MPC

In this section, we use the results in [3] to approximate the chance constraints that appear in the proposed formulations (7.6) and (7.7). We then develop a tractable methodology to reformulate the proposed robust formulations. The approximation is done in a way to provide a feasible solution for all scenarios of the uncertainty realizations with probabilistic guarantees. In the first step, a bounded set that contains the uncertainty realizations with a specific probability of violations is constructed. We then formulate a robust optimization problem with respect to that set and show that the solution is guaranteed to be feasible for the initial chance constrained problems (7.6) and (7.7) with the desired level of confidence.

Randomization-Based Reformulation

Define $\mathcal{B}_i(\gamma)$ to be a bounded set of uncertainty realizations and we assume that it is an axis-aligned hyper-rectangle for each agent i. Note that the choice of a hyper-rectangle is not restrictive and any convex set with convex volume could have been chosen instead [7]. We parametrize $\mathcal{B}_i(\gamma) := \times_{t=0}^{N_t-1}[\underline{\gamma}_t, \overline{\gamma}_t]$ by $\gamma = (\underline{\gamma}, \overline{\gamma}) \in \mathbb{R}^{2N_t}$, where $\underline{\gamma} = (\underline{\gamma}_0, \ldots, \underline{\gamma}_{N_t-1}) \in \mathbb{R}^{N_t}$ and $\overline{\gamma} = (\overline{\gamma}_0, \ldots, \overline{\gamma}_{N_t-1}) \in \mathbb{R}^{N_t}$. Consider now the following chance-constrained optimization problem

$$\min_{\gamma} \quad \sum_{t=0}^{N_t-1} \overline{\gamma}_t - \underline{\gamma}_t \tag{7.9}$$
$$\text{s.t.} \quad \mathbb{P}(w_i \in \Delta \mid w_{i,t} \in [\underline{\gamma}_t, \overline{\gamma}_t], \ \forall t \in \{0, \ldots, N_t - 1\}) \geq 1 - \epsilon.$$

By construction the problem (7.9) is a convex program and we can apply the standard scenario approach to obtain a solution as follows.

$$\min_{\gamma} \quad \sum_{t=0}^{N_t-1} \overline{\gamma}_t - \underline{\gamma}_t \tag{7.10}$$
$$\text{s.t.} \quad w_{i,t}^j \in [\underline{\gamma}_t, \overline{\gamma}_t], \quad \begin{cases} \forall t \in \{0, \ldots, N_t - 1\} \\ \forall j \in \{1, \ldots, N_s\} \end{cases},$$

where N_s is the required number of scenarios (7.8) for each agent $i \in \{1, \ldots, N\}$ with $d = N_t N$. The optimal solution of (7.10) γ^* is a feasible solution for the problem (7.9) with confidence $1 - \beta$.

Determine $\mathcal{B}_i(\gamma^*)$ for all i and define $\mathfrak{B}^* := \{\mathcal{B}_1(\gamma^*), \ldots, \mathcal{B}_N(\gamma^*)\}$ and pose the robust counterpart of the problems (7.6)–(7.7) where $\mathbf{W} \in \mathfrak{B}^* \cap \Delta^N$. Note that the robust counterparts of (7.6)–(7.7) are not randomized programs and instead, they are finite-horizon robust mixed-integer linear problems where the constraints have to be satisfied for all values of the uncertainty inside $\mathfrak{B}^* \cap \Delta^N$. It is worth to mention that any feasible solution of the robust counterparts of (7.6)–(7.7) is a feasible solution for the problems (7.6) and (7.7) with at least confidence of $1 - \beta$.

The robust counterpart problems are tractable and equivalent to mixed-integer linear programs, since the uncertainty is bounded in a convex set [18]. It is shown in [18] that the robust problems are tractable and remain in the same class as the original problems, e.g. robust mixed-integer programs remain mixed-integer programs, for a certain class of uncertainty sets. This is achieved under the assumptions that the constraint functions are linear and homogeneous with respect to the uncertainty vector. In the sequel, we describe a tractable scheme for the robust counterparts of (7.6)–(7.7).

Tractable Robust Reformulation

Following the methodology outlined in the previous section, we first define $\boldsymbol{\gamma}^o :=$ $[\gamma_0^o, \gamma_1^o, \ldots, \gamma_{N_t-1}^o] \in \mathbb{R}^{NN_t}$ to be a vector whose elements are the middle points of the hyper-rectangle $\mathfrak{B}^*$ and is defined as $\boldsymbol{\gamma}^o = 0.5(\overline{\boldsymbol{\gamma}}^* + \underline{\boldsymbol{\gamma}}^*)$ and each element of $\boldsymbol{\gamma}^o$ represents a vector for all agents $i \in \{1, 2, \ldots, N\}$. Consider now the following tractable reformulations of the proposed robust counterpart of problems (7.6) and (7.7).

$$\min_{\mathbf{U},\mathbf{V}} \quad \mathbf{J}(\mathbf{U}) \tag{7.11a}$$

$$\text{s.t.} \quad \mathbf{A}X_0 + \mathbf{B}\mathbf{U} + \mathbf{H}\boldsymbol{\gamma}^o + \boldsymbol{\eta} \geq 0, \ X_0 = X(k), \tag{7.11b}$$

$$\mathbf{E}\mathbf{U} + \mathbf{F}\mathbf{V} + \mathbf{P} \leq 0, \tag{7.11c}$$

where $\boldsymbol{\eta} := [\eta_0, \eta_1, \ldots, \eta_{N_t-1}] \in \mathbb{R}^{NN_t}$ is a vector with each element $\eta_t \in \mathbb{R}^N$ denoting a bound for the worst-case uncertainty realizations at step t for all agents $i \in \{1, 2, \ldots, N\}$. We refer to Proposition 7.1 below that shows how to achieve this bound. We next present a tractable reformulation of the proposed robust counterpart of problem (7.7).

$$\min_{\overline{\mathbf{U}},\mathbf{G},\mathbf{V}} \quad \mathbf{J}(\mathbf{U}) \tag{7.12a}$$

$$\text{s.t.} \ \mathbf{A}X_0 + \mathbf{B}\mathbf{U} + \mathbf{H}\boldsymbol{\gamma}^o + \boldsymbol{\eta} \geq 0, \ X_0 = X(k), \tag{7.12b}$$

$$\mathbf{E}\mathbf{U} + \mathbf{F}\mathbf{V} + \mathbf{P} \leq 0, \tag{7.12c}$$

$$\mathbf{U} = \overline{\mathbf{U}} + \mathbf{G}\boldsymbol{\gamma}^o + \boldsymbol{\eta}_g, \tag{7.12d}$$

$$\eta_{g,t} \leq [\mathbf{G}(\overline{\boldsymbol{\gamma}}^* - \boldsymbol{\gamma}^o)]_t, \ \forall t \in \{0, 1, \ldots, N_t - 1\}, \tag{7.12e}$$

$$\eta_{g,t} \leq [\mathbf{G}(\underline{\boldsymbol{\gamma}}^* - \boldsymbol{\gamma}^o)]_t, \ \forall t \in \{0, 1, \ldots, N_t - 1\}, \tag{7.12f}$$

where $\boldsymbol{\eta}_g := [\eta_{g,0}, \eta_{g,1}, \ldots, \eta_{g,N_t-1}] \in \mathbb{R}^{NN_t}$ is a vector with each element $\eta_{g,t} \in \mathbb{R}^N$. The following proposition shows the link between the tractable problems (7.11) and (7.12), and the proposed formulations (7.6) and (7.7), respectively.

Proposition 7.1 *If the tractable problems (7.11) and (7.12) have an optimal solution, where $\boldsymbol{\eta}$ is obtained by solving the following problem*

$$\max_{\eta \in \mathbb{R}^{N N_t}} \eta$$

$$\text{s.t. } \eta_t \leq [\mathbf{H}(\overline{\pmb{\gamma}}^* - \pmb{\gamma}^o)]_t \, , \forall t \in \{0, 1, \ldots, N_t - 1\}, \tag{7.13}$$

$$\eta_t \leq [\mathbf{H}(\underline{\pmb{\gamma}}^* - \pmb{\gamma}^o)]_t \, , \forall t \in \{0, 1, \ldots, N_t - 1\},$$

then it is a feasible solution for the chance-constrained problems (7.6) and (7.7) with at least $1 - \beta$ *confidence level, respectively.*

Proof It is shown in [19, Proposition 1] that any feasible solution of the robust counterparts of (7.6)–(7.7) is a feasible solution of the initial chance-constrained problems (7.6) and (7.7), respectively. Therefore, we have to show that the proposed tractable problems (7.11) and (7.12) are equivalent with the robust counterparts of (7.6)–(7.7). Consider the following robust constraint,

$$0 \leq \mathbf{A}X_0 + \mathbf{B}U + \mathbf{H}W , \; \forall W \in \mathfrak{B}^* \cap \Delta^N ,$$

that can be written in an equivalent format using the linearity and homogeneity assumption of the constraint with respect to the uncertainty, leading to

$$0 \leq \mathbf{A}X_0 + \mathbf{B}U + \mathbf{H}(\pmb{\gamma}^o + \Delta\pmb{\gamma}) =$$
$$\mathbf{A}X_0 + \mathbf{B}U + \mathbf{H}\pmb{\gamma}^o + \mathbf{H}\Delta\pmb{\gamma} , \; \forall \Delta\pmb{\gamma} \in [\underline{\pmb{\gamma}}^* - \pmb{\gamma}^o, \pmb{\gamma}^o - \overline{\pmb{\gamma}}^*] .$$

We need to introduce the vector $\pmb{\eta} := [\eta_0, \eta_1, \ldots, \eta_{N_t-1}] \in \mathbb{R}^{N N_t}$ with each element $\eta_t \in \mathbb{R}^N$ representing a bound for $\mathbf{H}\Delta\pmb{\gamma}$. Consider now the worst-case uncertainty realizations to be $(\overline{\pmb{\gamma}}^* - \pmb{\gamma}^o)$ and $(\underline{\pmb{\gamma}}^* - \pmb{\gamma}^o)$. We pose the problem (7.13) to find bound $\pmb{\eta}$. Using this bound leads to

$$0 \leq \mathbf{A}X_0 + \mathbf{B}U + \mathbf{H}\pmb{\gamma}^o + \pmb{\eta}$$
$$\leq \mathbf{A}X_0 + \mathbf{B}U + \mathbf{H}\pmb{\gamma}^o + \mathbf{H}\Delta\pmb{\gamma} = \mathbf{A}X_0 + \mathbf{B}U + \mathbf{H}(\pmb{\gamma}^o + \Delta\pmb{\gamma}).$$

The proof is completed. ∎

Remark 7.1 Note that we can use the same approach by introducing $\eta_{g,t}$ to be the worst-case superposition of the uncertainty realizations with the following constraints:

$$\eta_{g,t} \leq \sum_{j=0}^{t-1} \theta_{t,j} [(\overline{\pmb{\gamma}}^* - \pmb{\gamma}^o)]_j \, , \; \eta_{g,t} \leq \sum_{j=0}^{t-1} \theta_{t,j} [(\underline{\pmb{\gamma}}^* - \pmb{\gamma}^o)]_j .$$

Robust Randomized Model Predictive Control

The proposed procedure of a robust randomized MPC is summarized in Algorithm 4. We compare our proposed methodology to illustrate its performance against a hybrid approach as a benchmark, where the generating unit status problem is solved deterministically, meaning that we initialize $\pmb{\gamma}^o \equiv \mathbf{W}^{\text{forecast}}$, $\pmb{\eta} \equiv 0$ in (7.11), (7.12) with the forecast value of the energy demand and solve the deterministic variant of the problems. At the next step, we fix the on-off status of the generating units (and also the startup cost and auxiliary variables) to the binary vector computed by the previous deterministic program, and formulate a stochastic production planning problem. We refer to this as the Benchmark approach and the steps are summarized in Algorithm 5.

Algorithm 4 Robust Randomized MPC

1: Fix $X_0 = X(k)$, $\epsilon \in (0, 1)$, $\beta \in (0, 1)$ ▷ the initial (current) state measurement, level of constraint violations and confidence level of the agents, respectively.
2: Generate N_s scenarios (7.8) with $d = 2N_t N$ and establish $\mathfrak{B}^*$ by solving the optimization problem (7.10).
   Open-loop
3: Solve (7.11) and determine an optimal solution $\mathbf{U}^*$, $\mathbf{V}^*$. Apply the first optimal solution $U(k) := U_0^\star$, $V(k) := V_0^\star$ to the STG agents.
   Affine Uncertainty Feedback
4: Solve (7.12) and determine an optimal solution $\overline{\mathbf{U}}^*$, $\mathbf{G}^*$, $\mathbf{V}^*$. Apply the first optimal solution $U(k) := U_0^\star + [\mathbf{G}^* \boldsymbol{\gamma}^o + \boldsymbol{\eta}_g]_0$, $V(k) := V_0^\star$ to the STG agents.
5: $k \leftarrow k + 1$
6: Go to step 1.

Algorithm 5 Benchmark Approach

   Deterministic Generating Unit Status
1: Fix $X_0 = X(k)$ and $\boldsymbol{\gamma}^o \equiv \mathbf{W}^{\text{forecast}}$, $\boldsymbol{\eta} \equiv 0$ ▷ the initial (current) state measurement and no heat demand prediction error.
   Open-loop
2: Solve (7.11) and determine an optimal solution $\mathbf{V}^*_{\text{OLP}}$.
   Affine Uncertainty Feedback
3: Solve (7.12) and determine an optimal solution $\mathbf{V}^*_{\text{AUF}}$.
   Stochastic Production Planning
4: Fix $\epsilon \in (0, 1)$, $\beta \in (0, 1)$ and consider $\boldsymbol{\gamma}^o \equiv \mathbf{W}^{(j)}$, $\boldsymbol{\eta} \equiv 0$, $j = 1, \ldots, N_s$ ▷ level of constraint violations and confidence, respectively.
   Open-loop
5: Generate N_s scenarios (7.8) with $d = N_u N_t$.
6: Fix $V(k) = V^*_{\text{OLP,0}} \rightarrow$ Solve (7.11) and determine an optimal solution $\mathbf{U}^*$. Apply the first optimal solution $U(k) := U_0^\star$ to the STG agents. Go to step 2.
   Affine Uncertainty Feedback
7: Generate N_s scenarios (7.8) with $d = N_t(Nu + N_w N(N_t - 1)/2)$.
8: Fix $V(k) = V^*_{\text{AUF,0}}$ and solve (7.12) and determine an optimal solution $\overline{\mathbf{U}}^*$, $\mathbf{G}^*$. Apply the first optimal solution $U(k) := \overline{U}_0^\star + [\mathbf{G}^* \boldsymbol{\gamma}^o + \boldsymbol{\eta}_g]_0$ to the STG agents.
9: $k \leftarrow k + 1$
10: Go to step 1.

Remark 7.2 The proposed framework solves a stochastic mixed-integer program and it does not necessarily lead to a less conservative approach than the direct scenario approach [14], due to the fact that the number of required scenarios (7.8) is a function of the dimension of the decision variables. The decision variable size in our framework is proportional to the uncertainty dimension and in case of high uncertainty dimension, the advantage of our solution comes at the expense of a more conservative performance.

7.4 Distributed Stochastic MPC

In this section, we formulate two distributed approaches that are based on the alternating direction method of multipliers (ADMM) [20] to solve the robust randomized MPC (Algorithm 4). Due to the existence of the balance constraints (7.4h), problem (7.11) is not trivially separable. Therefore, we decompose the problem by considering the dual problem that is associated with problem (7.11). Furthermore, the ADMM approach is considered since the cost function is not strictly convex. In this section we present two ADMM formulations, which are a fully distributed scheme and a distributed scheme with a coordinator. Note that due to space constraints, we only address the distributed formulation of (7.11). However, problem (7.12) can also be solved in a distributed fashion with the same approach. Based on the proposed distributed formulations, Steps 3 and 4 in Algorithm 4 can be done in distributed manner.

Fully Distributed Scheme

Problem (7.11) can be expressed in a compact form as

$$\min_{\{\tilde{u}_i, \tilde{v}_i\}_{i=1}^N} \sum_{i=1}^N J_i(\tilde{u}_i) \tag{7.14a}$$

$$\text{subject to} \quad \tilde{u}_i \in \mathcal{L}_{u,i}, \quad \tilde{v}_i \in \mathcal{L}_{v,i}, \tag{7.14b}$$

$$\tilde{u}_i^a + \sum_{j \in N_{-i}} G_{ij} \tilde{u}_j^a = 0, \quad \forall i \in \{1, \dots, N\}, \tag{7.14c}$$

where for each agent i, $J_i(\tilde{u}_i) = \sum_{t=0}^{N_t-1} J_i(u_{i,t})$, $\tilde{u}_i = \left[u_{i,0}^\top, \dots, u_{i,N_t-1}^\top\right]^\top$, $\tilde{v}_i = \left[v_{i,0}^\top, \dots, v_{i,N_t-1}^\top\right]^\top$, and $\tilde{u}_i^a = \left[u_{i,0}^{a\top}, \dots, u_{i,N_t-1}^{a\top}\right]^\top$. Furthermore, $\mathcal{L}_{u,i}$ and $\mathcal{L}_{v,i}$ are the sets defined by the local constraints, i.e., (7.4b)–(7.4g), (7.4i)–(7.4l). Additionally, (7.14c) represents the coupling constraints, where G_{ij}, for each $j \in N_{-i}$, is defined appropriately according to (7.4h). We therefore can reformulate the problem as:

$$\min_{\{\tilde{u}_i, \tilde{y}_i, \tilde{v}_i\}_{i=1}^N} \sum_{i=1}^N J_i(\tilde{u}_i) + \psi_i(\tilde{y}_i) \tag{7.15a}$$

$$\text{subject to} \quad \tilde{u}_i \in \mathcal{L}_{u,i}, \quad \tilde{v}_i \in \mathcal{L}_{v,i}, \tag{7.15b}$$

$$\tilde{u}_i^a - \tilde{y}_i = 0, \quad \forall i \in \{1, \dots, N\}, \tag{7.15c}$$

where $\boldsymbol{\psi}_i(\tilde{\boldsymbol{y}}_i)$ corresponds to a convex indicator function such that $\psi_i(\tilde{\boldsymbol{y}}_i) = 0$ if $\tilde{\boldsymbol{y}}_i = -\sum_{j \in N_{-i}} \mathbf{G}_{ij} \tilde{\boldsymbol{u}}_j^a$ and $\psi_i(\tilde{\boldsymbol{y}}_i) = +\infty$ otherwise. Moreover, note that $\tilde{\boldsymbol{y}}_i = [\boldsymbol{y}_{i,0}^\top, \ldots, \boldsymbol{y}_{i,N_t-1}^\top]^\top \in \mathbb{R}^{N_t|N_{-i}|}$ for all $i \in \{1, \ldots, N\}$, are auxiliary variables. Consider the augmented Lagrangian of this problem as follows:

$$L_\rho = \sum_{i=1}^N \left(J_i(\tilde{\boldsymbol{u}}_i) + \boldsymbol{\psi}_i(\tilde{\boldsymbol{y}}_i) + \tilde{\boldsymbol{\lambda}}_i^\top (\tilde{\boldsymbol{u}}_i^a - \tilde{\boldsymbol{y}}_i) + \frac{\rho}{2} \|\tilde{\boldsymbol{u}}_i^a - \tilde{\boldsymbol{y}}_i\|_2^2 \right)$$

$$= \sum_{i=1}^N L_{\rho,i}(\tilde{\boldsymbol{u}}_i, \tilde{\boldsymbol{y}}_i, \tilde{\boldsymbol{\lambda}}_i).$$

where $\tilde{\boldsymbol{\lambda}}_i = [\boldsymbol{\lambda}_{i,0}^\top, \ldots, \boldsymbol{\lambda}_{i,N_t-1}^\top]^\top \in \mathbb{R}^{N_t|N_{-i}|}$ for all $i \in \{1, \ldots, N\}$ are the Lagrange multipliers and $\rho > 0$ is the penalty parameter. Hence, the augmented dual problem associated with Problem (7.15) is

$$\max_{\{\tilde{\boldsymbol{\lambda}}_i\}_{i=1}^N} \min_{\{\tilde{\boldsymbol{u}}_i, \tilde{\boldsymbol{y}}_i, \tilde{\boldsymbol{v}}_i\}_{i=1}^N} \sum_{i=1}^N L_{\rho,i}(\tilde{\boldsymbol{u}}_i, \tilde{\boldsymbol{y}}_i, \tilde{\boldsymbol{\lambda}}_i) \tag{7.16a}$$

$$\text{subject to} \quad \tilde{\boldsymbol{u}}_i \in \mathcal{L}_{u,i}, \quad \tilde{\boldsymbol{v}}_i \in \mathcal{L}_{v,i}, \quad \forall i \in \{1, \ldots, N\}, \tag{7.16b}$$

$$\tilde{\boldsymbol{y}}_i = -\sum_{j \in N_{-i}} \mathbf{G}_{ij} \tilde{\boldsymbol{u}}_j^a, \quad \forall i \in \{1, \ldots, N\}. \tag{7.16c}$$

We are now in a position to provide the ADMM steps that solves this problem in an iterative fashion:

1. Updating $\tilde{\boldsymbol{u}}_i$ and $\tilde{\boldsymbol{v}}_i$ for all $i \in \{1, \ldots, N\}$:

$$\{\tilde{\boldsymbol{u}}_i^{(q+1)}, \tilde{\boldsymbol{v}}_i^{(q+1)}\} \in \operatorname*{argmin}_{\tilde{\boldsymbol{u}}_i, \tilde{\boldsymbol{v}}_i} \quad L_{\rho,i}(\tilde{\boldsymbol{u}}_i, \tilde{\boldsymbol{y}}_i^{(q)}, \tilde{\boldsymbol{\lambda}}_i^{(q)})$$

$$\text{subject to} \quad \tilde{\boldsymbol{u}}_i \in \mathcal{L}_{u,i}, \quad \tilde{\boldsymbol{v}}_i \in \mathcal{L}_{v,i}.$$

2. Sending $\tilde{\boldsymbol{u}}_i^{a(q+1)}$ to the neighbors, $j \in N_{-i}$, for all $i \in \{1, \ldots, N\}$.
3. Receiving $\tilde{\boldsymbol{u}}_j^{a(q+1)}$ from the neighbors, $j \in N_{-i}$, for all $i \in \{1, \ldots, N\}$.
4. Updating $\tilde{\boldsymbol{y}}_i^{(q+1)}$ for all $i \in \{1, \ldots, N\}$:

$$\tilde{\boldsymbol{y}}_i^{(q+1)} = \operatorname*{argmin}_{\tilde{\boldsymbol{y}}_i} \quad L_{\rho,i}(\tilde{\boldsymbol{u}}_i^{(q+1)}, \tilde{\boldsymbol{y}}_i, \tilde{\boldsymbol{\lambda}}_i^{(q)})$$

$$\text{subject to} \quad \tilde{\boldsymbol{y}}_i = -\sum_{j \in N_{-i}} \mathbf{G}_{ij} \tilde{\boldsymbol{u}}_j^{a(q+1)},$$

which implies

$$\tilde{\boldsymbol{y}}_i^{(q+1)} = -\sum_{j \in N_{-i}} \mathbf{G}_{ij} \tilde{\boldsymbol{u}}_j^{a(q+1)}. \tag{7.17}$$

5. Updating $\tilde{\lambda}_i$ for all $i \in \{1, \ldots, N\}$ via a gradient method:

$$\tilde{\lambda}_i^{(q+1)} = \tilde{\lambda}_i^{(q)} + \rho \left(\tilde{u}_i^{a(q+1)} - \tilde{y}_i^{(q+1)} \right),$$

where $\rho > 0$.

The algorithm stops when

$$\left\| \begin{bmatrix} \tilde{u}_1^{a(q)} - \tilde{y}_1^{(q)} \\ \vdots \\ \tilde{u}_N^{a(q)} - \tilde{y}_N^{(q)} \end{bmatrix} \right\|_2 < \nu,$$

for a small $\nu > 0$. Note that the steps of updating $\tilde{y}_i$ and $\tilde{\lambda}_i$ are fully distributed among the agents. Moreover, not all decision variables but only $\tilde{u}_i^a$, which contains $h_{ex,t}^{ij}$, needs to be communicated between the agents.

One issue of an iterative algorithm, such as this method, is that it possibly requires a large number of iterations before the stopping criterion is met. In this regard, we apply warm start to reduce the number of iterations. It is done by using the solutions obtained in the previous sampling time since it often gives a good enough approximation [20]. For instance, consider that $\tilde{y}_i^{(q)} = \tilde{y}_i$ and $\tilde{\lambda}_i^{(q)} = \tilde{z}_i$ are the solutions obtained at the last iteration (q) at sampling time t. We initialize $\tilde{y}_i^{(0)}$ and $\tilde{\lambda}_i^{(0)}$ for the next sampling time as $\tilde{y}_i^{(0)} = \begin{bmatrix} \tilde{y}_i^{\top}(2 : N_t - 1) & \mathbf{0}_{1 \times |N_{-i}|} \end{bmatrix}^{\top}$ and $\tilde{\lambda}_i^{(0)} = \begin{bmatrix} \tilde{z}_i^{\top}(2 : N_t - 1) & \mathbf{0}_{1 \times |N_{-i}|} \end{bmatrix}^{\top}$.

Distributed Scheme with Coordination

The second ADMM method is formulated by perceiving the problem as an optimal exchange problem [20]. In this regard, we can consider restating problem (7.11) as follows:

$$\min_{\{\tilde{u}_i, \tilde{v}_i\}_{i=1}^N} \quad \sum_{i=1}^N J_i(\tilde{u}_i) \tag{7.18a}$$

$$\text{subject to} \quad \tilde{u}_i \in \mathcal{L}_{u,i}, \quad \tilde{v}_i \in \mathcal{L}_{v,i}, \tag{7.18b}$$

$$\sum_{i=1}^N \mathbf{K}_i \tilde{u}_i^a = 0, \tag{7.18c}$$

in which (7.18c) represents the balance constraints (7.4h). This formulation is different from 7.14 since there is only one global coupling constraint (7.18c) instead of N coupling constraints (7.14c). We can then follow the unscaled form of ADMM for such problems as provided in [20] that consists of the following iterations:

1. Updating of $\tilde{\boldsymbol{u}}_i$ and $\tilde{\boldsymbol{v}}_i$:

$$\{\tilde{\boldsymbol{u}}_i^{(q+1)}, \tilde{\boldsymbol{v}}_i^{(q+1)}\} \in \underset{\tilde{\boldsymbol{u}}_i, \tilde{\boldsymbol{v}}_i}{\operatorname{argmin}} \left\{ J_i(\tilde{\boldsymbol{u}}_i) + \boldsymbol{\lambda}^{(q)\top} \left(\mathbf{K}_i \tilde{\boldsymbol{u}}_i^{a(q)} \right) \right.$$
$$\left. + \frac{\rho}{2} \left\| \mathbf{K}_i \tilde{\boldsymbol{u}}_i^{a(q)} + \left(\tilde{\boldsymbol{y}}^{(q)} - \mathbf{K}_i \tilde{\boldsymbol{u}}_i^{a(q)} \right) \right\|_2^2 \right\}$$
$$\text{subject to} \quad \tilde{\boldsymbol{u}}_i \in \mathcal{L}_{u,i}, \quad \tilde{\boldsymbol{v}}_i \in \mathcal{L}_{v,i},$$

2. Updating of $\tilde{\boldsymbol{y}}$:

$$\tilde{\boldsymbol{y}}^{(q+1)} = \sum_{i=1}^{N} \mathbf{K}_i \tilde{\boldsymbol{u}}_i^{a(q+1)}.$$

3. Updating of $\tilde{\boldsymbol{\lambda}}$:

$$\tilde{\boldsymbol{\lambda}}^{(q+1)} = \tilde{\boldsymbol{\lambda}}^{(q)} + \rho \tilde{\boldsymbol{y}}^{(q+1)}.$$

In this approach, although the optimization problem is solved in a distributed fashion, the process of updating the auxiliary decision variable $\tilde{\boldsymbol{y}}$ and the Lagrange multiplier $\tilde{\boldsymbol{\lambda}}$ requires a coordinator that receives the decision of $\tilde{\boldsymbol{u}}_i^a$ at each iteration from all agents.

7.5 Numerical Study

We carried out Monte Carlo simulations and a comparison with the Benchmark Algorithm 5 to illustrate the performance of our proposed Algorithm 4 (robust randomized MPC) for both open-loop and closed-loop formulations. All optimization problems were solved using the solver BNB via the MATLAB interface YALMIP [21].

Simulation Setup

In this simulation study we consider a small thermal grid with three agents as an example. Figure 7.1 depicts the connections between each agent and their local components. Each agent has a micro-CHP, a boiler and a thermal storage. In the proposed model the difference between the level of thermal storage and local thermal energy demand (imbalance errors) is defined as the state of the local agent. The thermal storage level of agent one, two and three are presented in Fig. 7.1 using h_s^1, h_s^2, h_s^3, respectively. There are also three lines between agents indicating that thermal energy exchange is possible. We assume that an external party is available for all agents to provide thermal energy.

The proposed Algorithm 4 and the Benchmark approach are applied to the example provided in Fig. 7.1 with $N = 3$. We solve a day-ahead production planning problem for an uncertain thermal grid with $N_t = 24$ and hourly steps. It is assumed that the up and down capacity of decreasing and increasing electrical power are $p_g^{\text{up}} = p_g^{\text{down}} = p_g^{\text{max}}/3$ and the minimum time for a change of production unit status ($\Delta t_{\text{up}}, \Delta t_{\text{down}}$) is 2 h. Table 7.1 contains all parameters that are considered for the

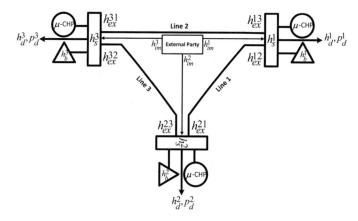

Fig. 7.1 Three-agent (households, greenhouses) thermal grid example. Each agent has a μ−CHP, a boiler and a thermal storage. h_s^1, h_s^2, h_s^3 are related to the local thermal storage in agent one, two and three, respectively. There are also three lines between agents indicating that thermal energy exchange is possible. It is considered to have an external party available for all agents to provide thermal energy

Table 7.1 Parameters with their symbols and values

Parameter	Value	Unit
p_g^{max}, h_g^{max}, p_g^{min}, h_g^{min}	120.0, 120.0, 0.0, 0.0	[KW]
h_b^{max}, h_{im}^{max}, h_b^{min}, h_{im}^{min}	120.0, 120.0, 0.0, 0.0	[KW]
h_{ex}^{max}, h_{ex}^{min}, $h_{s,0}^i$	20.0, −20.0, 10.0	[KW]
η_{CHP}, η_h, η_p	0.25, 0.7, 0.3	−
η_s, η_b, α_{ij}	0.85, 1.0, 0.25	−
c_{gas}, c_{up}, c_{dp}, c_b, c_{im}	45.0, 0.0, 100.0, 45.0, 300.0	e
Λ^{su} (micro-CHP, boiler)	diag(60.0, 120.0)	e
ϵ, β	0.1, 0.0001	−

example in Fig. 7.1. In order to generate scenarios for the thermal energy demand error, we used a Markov chain based model (we refer the reader to [22] for more details). Moreover, we consider to have different forecast of energy demand profiles for each agent. We construct scenarios of uncertain demand profiles assuming that the realization changes randomly to represent historical uncertain demand data.

Simulation Results: Centralized Stochastic MPC

Figure 7.2 shows imbalance error trajectories for different agents. Due to the definition of the imbalance error in Eq. (7.1), our goal is to minimize these errors. This means that the requested thermal energy demand is provided for each agent at each step with the desired level of violation ϵ as in Eq. (7.4l). The initial value for the storage level in each agent is considered to be 10 [KW]. In Fig. 7.2 the 'blue', 'red' and 'green' lines are related to the imbalance error profiles (x_1, x_2, x_3) in the first,

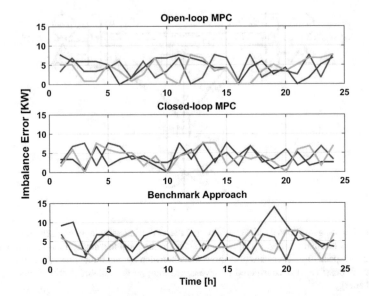

Fig. 7.2 Imbalance error trajectories. 'Blue' lines show the imbalance error x_1 in the first agent and x_2 the imbalance error in the second agent is shown by 'Red' lines. 'Green' lines represent the imbalance error x_3 in the third agent. The first, second and third sub-figures are related to the results of open-loop MPC, closed-loop MPC considering affine uncertainty feedback, and the Benchmark approach, respectively

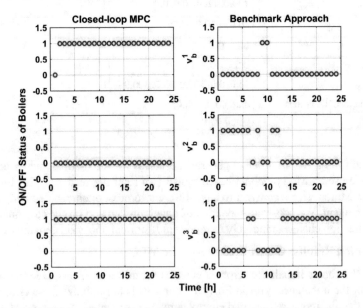

Fig. 7.3 ON/OFF status of boilers. The first, second and third sub-figures are related to agent 1, 2, and 3, respectively. The sub-figures on the left are the results of closed-loop MPC considering affine uncertainty feedback, and the sub-figures on the right are the results of the Benchmark approach

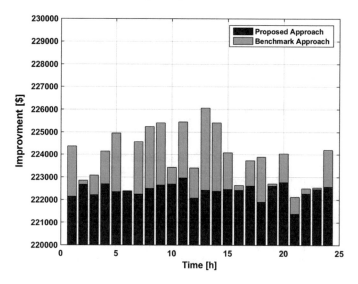

Fig. 7.4 Relative cost improvement (expressed in $) of the closed-loop MPC approach (affine uncertainty feedback) over the Benchmark approach

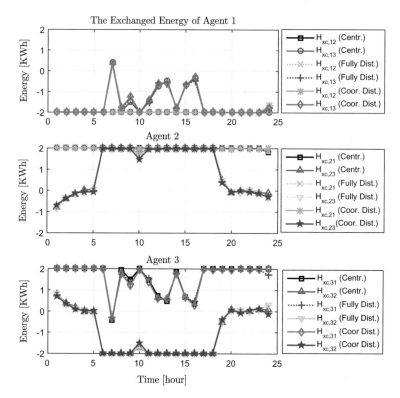

Fig. 7.5 The thermal energy exchanged between agents

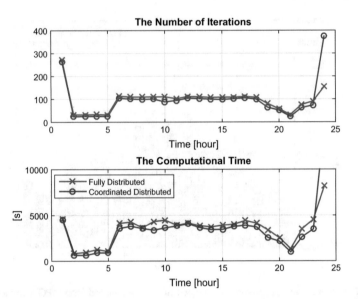

Fig. 7.6 The number of iterations and the computational time required by the distributed approaches at each sampling time

second and third agent. The top sub-figure shows the result of optimizing directly over input sequences in each sampling time (open-loop MPC), the middle depicts the result of closed-loop MPC considering affine uncertainty feedback, and the last sub-figure shows the result of the Benchmark approach. In Fig. 7.3 the ON/OFF status of boilers at each sampling time using blue for the first, second and third agent are shown, respectively. The left-hand-side sub-figures are related to the closed-loop MPC considering affine uncertainty feedback, and the-right-hand side ones show the results of the Benchmark approach. In Fig. 7.4 the relative cost improvement between the cost generated by the closed-loop MPC considering affine uncertainty feedback and the cost generated by the Benchmark approach in each sampling time is shown.

The proposed Algorithm 4 offers a better working plan for production units as well as an hourly-based production cost improvement compared to the Benchmark approach. The improvement in terms of cost is due to the scheduling flexibility offered by the proposed algorithm, where the binary variables are solved together with the production planning problem, allowing us to identify more optimal working status for production units.

Simulation Results: Distributed Stochastic MPC

We also compare the performance of the distributed approaches with the centralized one. In summary, the distributed approaches are able to find quite similar solutions to the centralized one. Moreover, Fig. 7.5 shows the satisfaction of the balance constraints in some agents when the distributed approaches are employed. Furthermore, Fig. 7.6 shows the number of iterations and the computational time required by the distributed approaches to solve the problem at each sampling time. It can be seen that both distributed methods have similar number of iterations as well as computational times.

7.6 Concluding Remarks

This work formulated an optimization problem for a day-ahead prediction plan of smart thermal grids with uncertain local demands. Smart thermal grids refer to energy networks whose main goal is to provide and distribute thermal energy among their local agents. This formulation leads to a multistage chance-constrained mixed-integer linear program. We proposed a unified framework, namely a robust randomized MPC approach to solve such a problem, while providing a-priori guarantees for the chance constraint fulfillment. Additionally, we also propose to apply a distributed optimization method based on ADMM to solve the robust randomized MPC program. Our current work focuses on incorporating aquifer thermal energy storage (ATES) systems in the developed framework.

References

1. H. Lund, S. Werner, R. Wiltshire, S. Svendsen, J.E. Thorsen, F. Hvelplund, B.V. Mathiesen, 4th generation district heating (4GDH): integrating smart thermal grids into future sustainable energy systems. Energy **68**, 1–11 (2014)
2. S. Grijalva, M.U. Tariq, Prosumer-based smart grid architecture enables a flat, sustainable electricity industry, in *Innovative Smart Grid Technologies Conference* (IEEE, 2011), pp. 1–6
3. A. Bemporad, M. Morari, Robust model predictive control: a survey, in *Robustness in Identification and Control* (Springer, 1999), pp. 207–226
4. M.V. Kothare, V. Balakrishnan, M. Morari, Robust constrained model predictive control using linear matrix inequalities. Automatica **32**(10), 1361–1379 (1996)
5. P. Scokaert, D. Mayne, Min-max feedback model predictive control for constrained linear systems. Trans. Autom. Control **43**(8), 1136–1142 (1998)
6. M.C. Campi, S. Garatti, M. Prandini, The scenario approach for systems and control design. Ann. Rev. Control **33**(2), 149–157 (2009)
7. K. Margellos, V. Rostampour, M. Vrakopoulou, M. Prandini, G. Andersson, J. Lygeros, Stochastic unit commitment and reserve scheduling: a tractable formulation with probabilistic certificates, in *European Control Conference (ECC)* (IEEE, 2013), pp. 2513–2518
8. X. Zhang, K. Margellos, P. Goulart, J. Lygeros, Stochastic model predictive control using a combination of randomized and robust optimization, in *Conference on Decision and Control* (IEEE, 2013), pp. 7740–7745
9. P.D. Christofides, R. Scattolini, D. Muñoz de la Peña, J. Liu, Distributed model predictive control: a tutorial review and future research directions. Comput. Chem. Eng. **51**, 21–41 (2013)
10. V. Rostampour, T. Keviczky, Robust randomized model predictive control for energy balance in smart thermal grids, in *European Control Conference (ECC)* (IEEE, 2016), pp. 1201–1208
11. W. Ananduta, Distributed energy management in smart thermal grids with uncertain demands, M.Sc. dissertation Delft University of Technology, The Netherlands (2016)
12. M. Prandini, S. Garatti, J. Lygeros, A randomized approach to stochastic model predictive control, in *Conference on Decision and Control* (IEEE, 2012), pp. 7315–7320

13. P. Goulart, E. Kerrigan, J. Maciejowski, Optimization over state feedback policies for robust control with constraints. Automatica **42**(4), 523–533 (2006)
14. G.C. Calafiore, M.C. Campi, The scenario approach to robust control design. Trans. Autom. Control **51**(5), 742–753 (2006)
15. M.C. Campi, S. Garatti, The exact feasibility of randomized solutions of uncertain convex programs. SIAM J. Optim. **19**(3), 1211–1230 (2008)
16. V. Rostampour, K. Margellos, M. Vrakopoulou, M. Prandini, G. Andersson, J. Lygeros, Reserve requirements in ac power systems with uncertain generation, in *Innovative Smart Grid Technologies Europe (ISGT EUROPE)* (IEEE, 2013), pp. 1–5
17. P. Mohajerin Esfahani, T. Sutter, J. Lygeros, Performance bounds for the scenario approach and an extension to a class of non-convex programs. IEEE Trans. Autom. Control **60**(1), 46–58 (2015)
18. D. Bertsimas, M. Sim, Tractable approximations to robust conic optimization problems. Math. Program. **107**(1–2), 5–36 (2006)
19. K. Margellos, P. Goulart, J. Lygeros, On the road between robust optimization and the scenario approach for chance constrained optimization problems. Trans. Autom. Control **59**(8), 2258–2263 (2014)
20. S. Boyd, N. Parikh, B.P.E. Chu, J. Eckstein, Distributed optimization and statistical learning via the alternating direction method of multipliers. Found. Trends Mach. Learn. **3**(1), 1–122 (2010)
21. J. Löfberg, Yalmip: a toolbox for modeling and optimization in matlab, in *International Symposium on Computer Aided Control Systems Design* (IEEE, 2004), pp. 284–289
22. V. Rostampour, P. Mohajerin Esfahani, T. Keviczky, Stochastic nonlinear model predictive control of an uncertain batch polymerization reactor, in *IFAC Conference on Nonlinear Model Predictive Control (NMPC)*, vol. 48, No. 23 (Elsevier, 2015), pp. 540–545

Part V
Vehicle to Grid Technologies and Institutional Arrangements

Chapter 8
Fuel Cell Electric Vehicle-to-Grid Feasibility: A Technical Analysis of Aggregated Units Offering Frequency Reserves

C. B. Robledo, M. J. Poorte, H. H. M. Mathijssen, R. A. C. van der Veen and A. J. M. van Wijk

Abstract Fuel Cell Electric Vehicles (FCEVs) in combination with green hydrogen (obtained from renewable sources), could make a significant contribution in decarbonizing the European transport sector, and thus help achieve the ambitious climate goals. However, most vehicles are parked for about 95% of their life time. This work proposes the more efficient use of these vehicles by providing vehicle-to-grid (V2G) services achieving the integration of the transport and energy systems. The aim of this work is to determine the technical and financial potential value that FCEVs could have by providing frequency reserves. Experiments were carried out with a Hyundai ix35 FCEV that was adapted with a power output socket so it can operate in V2G when parked, delivering maximum 10 kW direct current power. Results show that both power sources in the fuel cell electric vehicle, which are the fuel cell stack and the battery, can react in the order of milliseconds and thus are suitable to offer fast frequency reserves. The challenge lays in the communication between the car and the party that sends the signal for the activation of the frequency reserves. As one

C. B. Robledo (✉)
Process and Energy, 3mE, Delft University of Technology, Leeghwaterstraat 39, 2628CB Delft, The Netherlands
e-mail: c.b.robledo@tudelft.nl

M. J. Poorte
Talent voor Transitie, Bergstraat 35, 6811LC Arnhem, The Netherlands
e-mail: michelle.poorte@talentvoortransitie.nl

H. H. M. Mathijssen
TenneT, Utrechtseweg 310, 6812AR Arnhem, The Netherlands
e-mail: Henrie.Mathijssen@tennet.eu

R. A. C. van der Veen
CE Delft, Oude Delft 180, 2611HH Delft, The Netherlands
e-mail: veen@ce.nl

A. J. M. van Wijk
Process and Energy, 3mE, Delft University of Technology, Leeghwaterstraat 39, 2628CB Delft, The Netherlands
e-mail: a.j.m.vanwijk@tudelft.nl

© Springer Nature Switzerland AG 2019
P. Palensky et al. (eds.), *Intelligent Integrated Energy Systems*,
https://doi.org/10.1007/978-3-030-00057-8_8

unit does not provide enough power to be able to participate in the electricity market, a car park acting as aggregator of FCEVs was designed taking into account current technology developments. A carpark with a direct current microgrid, a hydrogen local network and only occupied by FCEVs was designed. A financial model was developed to evaluate the economic potential of the car park to participate in the electricity market providing frequency reserves. Results show that by using the fuel cells in the FCEVs in V2G, monetary benefits could be obtained when providing automated frequency restoration reserves (aFRR) upwards. Key parameters are found to be the investment costs, amount of vehicles available, hydrogen price and price of aFRR. With a car park of approximately 400 cars all year long available, payback times of 11.8 and 3.5 years were obtained taking into account worst and best case scenarios for a 15 year period analysis, respectively.

8.1   Introduction

In the process of liberalization of the electricity market in Europe that took place in the late 1990s, the electricity industry has become a competitive market, where different entities belonging to the Distribution System Operators (DSOs) and Transmission System Operators (TSOs) bear responsibility for the operation and security of the power system [1]. Users connected to the power system expect that the frequency and voltage stay close to their nominal values in order to ensure that their electronic devices will operate correctly. The frequency of the power system is used as an indicator of the (im)balance between the generation and the demand of the total active power in the system. Therefore, frequency control represents a crucial part of ancillary services [2]. Since the undergoing process of the electricity market and the increase of intermittent generation, the European power system has been exposed to high and persisting frequency deviations [1]. This is mainly due to the increase in the electricity production mix of wind and solar power, driven by the energy transition to a more sustainable power system [3, 4]. New and innovative technologies will have to be implemented in the current system in order to cope with the environmental needs and to secure constant electricity supply in the future energy system.

Electric vehicles (EVs) with vehicle-to-grid (V2G) technology have the capability to provide electricity to the grid, and have thus the potential to join the electricity market in providing ancillary services [5, 6]. Personal vehicles are utilized only 5% of their lifetime for transportation. The remaining time they can be used for a secondary function, like offering frequency reserves [7]. However, the individual participation of single EVs as electricity providers is not possible for mainly two reasons. First, the individual capacity is too small to participate in power system markets and second, the availability of one single EV is unpredictable from the power system's perspective as its main purpose is transportation. A larger fleet of vehicles could provide a solution by delivering a significant amount of reserves to the power system, while being more reliable. Such a mechanism to form clusters of smaller single units, that together could offer services, is called aggregation and is administered by the aggregator [8]. An aggregator acts as a mediator between the system operator and small-scale customers, enabling mutually beneficial coordination for EV owners and the power system [9]. The key role of the aggregator is to present the dispersed power units as a single cluster to the system operator.

There are mainly three types of EVs that could be used in V2G mode; plug-in hybrid (PHEV), battery (BEV) and fuel cell electric vehicles (FCEV) [10]. BEVs can store electricity and deliver it back to the grid when needed, whereas PHEVs and FCEVs can generate electricity from fuel stored on-board or fed into the car. We focus on FCEVs because they provide the only possibility to generate clean electricity, if hydrogen is produced from renewable energy sources. When parked, these cars can produce electricity in a clean and efficient way with only water and heat as by-products [6]. It is expected that future energy networks will allow higher level of interactions between electrical grid, thermal grind and fuel grid with hydrogen as the energy carrier [11]. Hydrogen can become an important energy carrier next to electricity, which can be transported by pipeline, with electrolysers and FCEVs becoming important units for power system balancing. FCEVs can use hydrogen for driving, and for power production when parked. Thus, hydrogen is a potential energy carrier to allow higher interactions between the different energy grids.

Technical capability and potential income generation of V2G services provided with FCEVs are not known yet but essential for the feasibility of FCEV aggregation providing frequency reserves. Therefore, in this chapter we seek to do so in two separate analyses. An experimental approach is used to test the dynamic response of a FCEV connected to the grid and delivering electricity at different power outputs. The second approach consists of the design and financial modeling of a car park, acting as aggregator, to provide several frequency control products in the energy market. We give a brief introduction to the frequency control system in Europe, and in particular on the Dutch system where the experiments were carried out. Also, fundamental aspects of V2G technology and hydrogen fuel cell electric vehicles are addressed.

8.1.1   Dutch Electric Power System

The Dutch electricity network is embedded in the synchronous area (SA) of Continental Europe, which is the largest one in the world and supplies over 400 million customers in 24 countries [12]. The most relevant frequency parameters and targets have been identified and set by the European Network of Transmission System Operators for Electricity (ENTSO-E) to guarantee a secure power system. Within the ENTSO-E network there are five SAs: Continental (CE), Nordic (NE), Baltic, British (BE), and Irish (IRE) Europe. The quality parameters are embedded in formal EU regulation in the System Operation Guideline (SO GL). In Europe, the nominal frequency has been set at 50 Hz. Frequency deviations from this value could lead to a complete blackout if not treated properly in time.

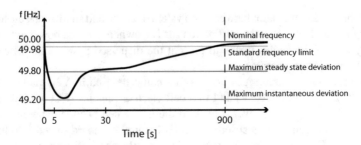

Fig. 8.1 Power system frequency along time after a power generation outage at time 0 [14]

8.1.1.1 Frequency Control

Figure 8.1 shows the dynamic behavior of the frequency after a power generation outage. At the moment of power outage (time = 0), a decrease of the frequency can be noticed. The maximum allowed deviation occurring after an incident is called the maximum instantaneous frequency deviation, which is equal to a deviation of 800 mHz compared to the nominal frequency. The maximum steady state frequency deviation is the limit that the frequency can reach without bringing the stability of the system in danger. The time to restore frequency after a power output indicates the maximum duration of a deviation until the frequency is restored to the standard frequency range (49.98–50.00 Hz) and is equal to 15 min. To bring the frequency back within the standard frequency range a control mechanism called Load Frequency Control (LFC) is established. The mechanism of LFC is a system that can activate power reserves in case of an imbalance. If there is a shortage of power, the frequency reserves can be activated and ramp up; if there is abundance, they can ramp down. In the European system, there are three types of reserves that contribute to the stabilization of the frequency: frequency containment reserves (FCR), frequency restoration reserves (FRR), and replacement reserves (RR). The RR are not used in the Dutch system and will therefore not be considered in this research.

When a disturbance takes place, first the FCR are activated. As a result, the deviation of the frequency stops and stabilizes at a value that differs from the nominal value. Then, FRR are activated to bring the frequency back to its nominal value. The time required to ramp up or down the reserves influences the time it takes to stabilize the frequency. Currently the full activation time (FAT), which is the time to activate the needed reserves, is 30 s for FCR. The FRR must be activated within 15 min (900 s). These time frames are indicated in Fig. 8.1.

The amount of FCR required per TSO is calculated each year based on the share of electricity that is produced and consumed within the operation area of that TSO and the total synchronous area. For the Dutch TSO TenneT, the amount of FCR has been approximately 100 MW in the past years. This reserve must be offered symmetrically, meaning that if 10 MW of FCR are offered, then the power source must be able to both ramp up and ramp down 10 MW.

Fig. 8.2 Timeline of the electricity markets. The dotted line indicates the actual moment of electricity transmission

The FRR is used to balance the power to its scheduled value. This process involves automatically instructed services (aFRR), previously known as secondary reserves, and manually instructed services (mFRR), known as tertiary reserves [13]. Currently the required amount of available aFRR in the Netherlands is 340 MW upward and 340 MW downward. These reserves are contracted separately. In this research mFRR, used for large-scale and expected long-lasting imbalances, will not be taken into consideration, as these reserves are subject to long-term contracts and are sporadically used. This makes it less interesting for the small-scale technologies to participate in this market.

8.1.1.2 Market Regulations

The Dutch electricity market can be divided in four submarkets based on the time until the actual electricity production and consumption takes place. Figure 8.2 shows an overview of these four markets in chronological order with respect to the moment of trade delimited as a dotted line. On the forward (or long-term) market the largest share of electricity is sold. This is done with contracts of years, quarters or months. The day-ahead market takes place one day before the actual transmission of electricity. On this market expected changes in production or consumption can be adjusted. This trade takes place by means of an auction. Bids can be done 24 h a day starting 45 days before the delivery day. The auction is open until 12.00 h of the day before transmission. The auction is cleared and a price per hour is determined; this is by the marginal price on the bid ladder [15]. The intraday market opens when the operators of TenneT have checked the balance on the grid, which is usually around 17.00 h of the day before the electricity transmission. It closes 5 min before the actual delivery. This trade also takes place in the form of an auction in blocks of one hour [16]. The fourth market is the balancing market, which is the entirety of institutional, commercial and operational arrangements that establish market-based management of balancing. This includes also the markets for frequency reserves. This market is spread over a larger time frame. FCR is traded before the electricity transmission. However for FRR, there is a bidding, activation and a settlement phase. The first phase occurs before the scheduled electricity is produced, the second phase takes place in real time and the settlement phase takes place after the electricity production.

8.1.1.3 Responsibilities

To manage the activation of the frequency reserves, a cascade of responsibilities, that begin at European level, has been established in the SO GL, which was drafted by ENTSO-E and is official EU regulation since 14 September 2017 [17].

To be able to balance supply and demand, the Dutch law attributes the administrative responsibility of forecasting generation and demand to the balance responsible parties (BRP). During the planning phase, the BRP provides energy-programs (E-programs) to the TSO a day before the day of delivery. An E-program contains the transactions of energy during an imbalance settlement period (ISP). After gate closure time, the BRP have the responsibility to accept the imbalance price in case they have deviated from the scheduled net energy injection or withdrawal [18].

The last relevant actors for the balancing mechanism are the balancing service providers (BSP). These provide the reserves required for balancing. The BSP offer their capacity for a given price on the specific markets for frequency reserves. For the different frequency reserves there are different activation mechanisms.

8.1.2 Fuel Cell Electric Vehicles and V2G

FCEVs store energy in molecular hydrogen H_2, which is fed into a fuel cell along with atmospheric oxygen, producing electricity with heat and water as by-products. Commercially available FCEVs use Proton Exchange Membrane Fuel Cells (PEMFCs) to convert hydrogen into electricity, have a High Voltage (HV) battery connected in parallel and store H_2 in gaseous form at 700 bar [19–22]. This electricity is used to drive the electric motor and also to charge the HV battery. For a moderate power demand, the vehicle feeds only from the fuel cell to power the traction motor. When the vehicle is moving uphill or accelerating, the traction motor is driven by the battery and fuel cell together [23].

Even before the term V2G was introduced in the literature, Kissock analyzed the potential of FCEVs to provide power for buildings and focused primarily on the ability of using waste heat for both space heating and to help operate an absorption cooling system for different buildings [24]. Three years later, Kempton et al. introduced the term V2G and examined the economic potential of using different types of EVs to produce power for buildings and the grid, as well as to provide ancillary services [10]. They concluded that FCEVs could compete in the peak power market but could not compete with baseload power. Lipman et al. assessed the economic benefits from a customer side of stationary fuel cells and FCEVs providing power for residential and commercial use. They concluded that FCEVs used for power in commercial "office building" can potentially supply electricity at competitive rates, in some cases producing significant annual benefits to vehicle owners while at the same time producing additional capacity to utility grids [25]. Also, the FCEVs have been considered to provide balancing power for smart city areas in a 100% renewable energy system [26]. All these relevant studies have analyzed the potential of FCEVs to provide power for different electricity markets theoretically, and in 2016 the main concept was tested in reality for the first time at the Green Village in The Netherland, the living lab of Delft University if Technology. A commercial FCEV was adapted and was connected to the national Dutch grid, delivering 9.5 kW AC power [27]. This was done within the Car as Power Plant project which entails a vision of an integrated sustainable energy and mobility system, based on renewable energy sources, hydrogen as an energy carrier and FCEVs [6]. A main business case related to the Car as Power Plant concept is that of the Car Park as Power Plant (CPPP), which consists of the aggregation of FCEVs parked in a car park into a 'virtual power plant' to produce electricity and feed it into the power grid [28]. This work elaborates on this existing concept and brings new insights on the use of FCEVs for frequency reserves.

8.2   Methodology

In this chapter we explain the methods employed to analyze the use of aggregated FCEVs to offer frequency reserves. First, the technical capability of a commercial available FCEV to ramp up and down according to FCR requirements was tested. Then, the results were taken as input in a model developed to simulate a car park aggregating FCEVs and taking into consideration the technical characteristics of the cars, relevant actors and the market regulations. Details of both methods are exposed in the following sub-sections.

8.2.1   Technical Evaluation of a FCEV to Provide Power to the Grid

The main objective of the experiment was to estimate the reaction time of the FCEV in V2G mode, which is a key component in estimating the FAT. The FAT stands for the time the balancing resource needs to ramp up from zero to the maximum power output or ramp down from the maximum power to zero. The two frequency reserves taken into consideration are FCR and aFRR. Considering that the FAT for FCR is shorter than that for aFRR, if the FCEV satisfies the requirements for the first, it also satisfies the latter. The FAT is composed of two parts, as shown in Eq. (8.1).

$$FAT = ReactionDelay + ReactionTime \qquad (8.1)$$

The reaction delay is the time related to the data communication, which is measured from the moment that a signal is sent by the BRP until the moment that the balancing resource actually reacts. The reaction time of the FAT is the time that the balancing resource actually adjusts the power output and reaches the point of full activation. As no actual aggregation, control mechanism and communication system is yet available for the CPPP concept, the reaction delay of this aggregation was not measured. During the experiment only the reaction time of the balancing resource was measured and the reaction delay was obtained from a V2G pilot with BEVs in The Netherlands, as this is equal to both type of EVs.

8.2.1.1   Experimental Setup

The tests were carried out at the Car as Power Plant set up located at The Green Village in The Netherlands, shown in Fig. 8.3. The Green Village is a unique platform that unites researchers, entrepreneurs, government and general public to help innovations get to large-scale application [29]. The experimental set-up consisted of mainly three components:

1. The modified Hyundai ix35 FCEV with V2G Direct Current (DC) outlet plug and data monitoring system on-board.
2. V2G unit with inverter inside, that converts DC power in the range of 300-400 V received from the FCEV into three-phase Alternating Current (AC) at 380 V.
3. A three-phase 380 V AC grid connection with galvanic isolation, including fuses and kWh meter.

The Hyundai ix35 FCEV has a 100 kW FC on board, connected to a Li-polymer battery in parallel, that has an energy capacity of 0.95 kWh and a maximum power output and input of 24 kW [30]. It is equipped with a large radiator and cooling

Fig. 8.3 Photo of the modified Hyundai ix35 FCEV connected to the grid through the V2G unit at The Green Village in The Netherlands

fan to maintain the temperature at a desirable operating level. In stationary mode the FCEV has less cooling capacity and cannot operate at the maximum power of 100 kW. For this reason the maximum power output in V2G mode was restricted to 10 kW DC that enables the on-board utilities of the FCEV to regulate the fuel cell's temperature [31]. More specifics about the technical and safety details of the set-up can be found in this reference [26].

8.2.1.2 Measurements

To determine the reaction time of the FCEV connected to the grid, the power output of the FCEV was measured while the inverter at the V2G unit was switched on and off. This was done to simulate when the car would be asked to deliver power by the aggregator in a future CPPP system. While the FCEV was connected to the V2G unit, turned on and not delivering power, it remained in idling condition. Once the inverter in the discharge unit was switched on, the FCEV actively started delivering power. By switching on and off the discharge unit, the FCEV was activated (ramp up) and deactivated (ramp down).

Two different power settings were tested. The first setting was for a power output of 10 kW DC from the car and the second one, a power output of 5 kW DC. As it can be seen in Fig. 8.4, the FCEV was allowed to idle for 2 min, followed by the delivery of power during 5 min. This cycle was repeated 10 times to improve the reliability of the measurements for both settings.

During the experiments, current and voltage was monitored and measured in the fuel cell and battery of the FCEV with a frequency of 5 Hz. The power values were calculated by multiplying the individual current and voltage values. The total DC power output of the FCEV in V2G was equal to the sum of the FC and the battery DC power output. With these values the power gradient ($\Delta P / \Delta t$) was calculated as

Fig. 8.4 Power input settings during the experiment with the FCEV delivering power in V2G mode

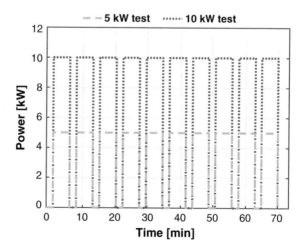

expressed in Eq. (8.2).

$$\frac{\Delta P}{\Delta t} = (P_{t+1} - P_t) \times f \tag{8.2}$$

where P_{t+1} is the power output of battery or FC at time t+1, P_t is the power output at time t, and f is the logging frequency. This was calculated for each cycle and for each component (battery and FC), when ramping up (switching on), as well as ramping down (switching off). The average of these measurements was used as indicator of the reaction time of the FCEV.

8.2.2 Financial Modeling of Car Park with FCEVs Providing Frequency Reserves

The overall scope of the financial analysis is to evaluate for which conditions the net present value (NPV) and the payback time of a CPPP offering frequency reserves are, respectively maximized and minimized. The model evaluates every 15 min of a year, all the combinations of the products that can be offered by the CPPP (electricity, FCR and aFRR) and the related energy that is used to offer these products. A time step of 15 min is chosen, as it is equal to the imbalance settlement period (ISP) of the Dutch power system. To calculate the NPV and the payback time four main steps were performed. First, the total power that can be offered by the car park was calculated. Then, the constraints related to the balancing market were implemented. Thirdly, the revenues and costs were calculated, and finally the yearly profit (calculated with the revenue and costs of the previous step) was used to obtain the NPV and payback time based on investment estimations. In the following two sub-sections the system design and model assumptions are explained.

8.2.2.1   System Design and Assumptions

In order for the concept of CPPP to be feasible, the fleet of FCEVs has to increase with respect to its current level. In 2016 only 31 vehicles were FCEVs within the total amount of 8,222,974 registered vehicles in The Netherlands [32]. We assume that the concept of a car park with FCEVs that operates as a power plant could become feasible from 2030 onwards. This is considering mainly three recent local developments:

1. In 2017, the new Dutch government presented its plans for the coming years and established that by 2030 all produced and sold cars must be emission free. In this category fall only battery and fuel cell electric vehicles.
2. The whitepaper presented by H_2 platform in May 2018 to be incorporated in the climate agreement in The Netherlands. The goal is to increase the amount

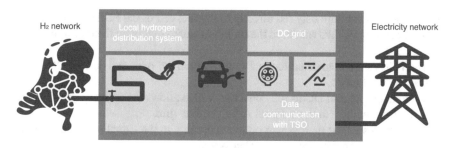

Fig. 8.5 Schematic representation of the CPPP system design with its main components; local hydrogen network, hydrogen dispensing, DC electrical grid in the car park, conversion to AC and data communication components

of FCEVs and Hydrogen fuel stations in The Netherlands to 375,000 and 216, respectively by 2030 [33].

3. EU targets to reduce the amount of CO_2 emitted per km for passenger cars from 130 g CO_2/km in 2015 to 95 g CO_2/km in 2021 and 70 g CO_2/km in 2025 [17].

The demand for hydrogen will rise and not only for its use in transportation, but also to satisfy the heating demand in residential and commercial buildings [34]. For higher demand of hydrogen, large-scale hydrogen production and low-pressure hydrogen pipelines for distribution will be more economical than local production and use of hydrogen [35].

In the future, a trend towards full electric is expected. But also an increase in energy sources providing DC power is expected [36]. The projected increase in the number of DC powered components for residential and industrial application, together with more distributed generation units that generate DC power, reveals that DC microgrid systems will soon be the right candidates for future energy systems [37, 38]. The advantages of low voltage DC grids are higher efficiency, material saving and longer lifespan [39]. Especially because battery and fuel cell in FCEVs, respectively store and produce DC electricity, a DC microgrid is preferred in the CPPP mid-century concept design.

Taking into account all these future developments, a system design for the CPPP was developed, which makes use of a local hydrogen distribution network, an internal DC microgrid and solely FCEV occupation. Figure 8.5 shows a scheme of the CPPP design.

The following system-level assumptions were applied to the CPPP model:

- The starting point is an existing car park to which several technical components are added.
- FCEVs are owned by the individual drivers, and the car park operator acts as the aggregator providing services to the energy market.
- The CPPP operator acquires hydrogen from the local low-pressure hydrogen distribution network.

- Hydrogen is continuously fed directly to the fuel cell of the FCEVs while parked and operating in V2G mode, so the on-board stored hydrogen meant for driving is not used.
- The FCEVs are connected to the DC grid of the car park and deliver DC power back in V2G mode.
- There is one central DC to AC converter, which allows the delivery of AC power at the right voltage to the national AC electricity grid.
- The vehicles adjust their power output according to the signal received from the TSO through the data communication system.

Capacity of the Car Park

The size of the car park is determined by the amount of parking places. More parking places per car park implies more potential capacity to offer to the power system. In the Netherlands there are a total of 210,000 parking places in car parks. This excludes the places near hospitals, parking territories and companies. These parking places are divided amongst circa 500 car parks [17]. This means that on average there are 420 parking places in one car park. This amount is considered as the maximum capacity of the car park under study. The maximum power that the car park can offer ($P_{V2G,max}$) is determined by the amount of occupied parking places (vehicles) times the maximum power output of a FCEV (P_{FCEV}) in stationary mode, as shown in Eq. (8.3).

$$P_{V2G,max} = Vehicles \times P_{FCEV} \qquad (8.3)$$

To evaluate the effect of the availability of cars on the energy services that the CPPP system could provide, two different occupation patterns were evaluated; A variable *city car park* occupation pattern based on real world data, and a constant *commercial car park*. The city car park represented a standard car park in The Netherlands, while the commercial car park represented those near airports, large convention centers, hotels and hospitals, which are always operating close to their maximum capacity. This occupation pattern is an almost constant value, approaching the maximum number of parking places. Figure 8.6 shows a plot of the amount of cars during a week that are present in both type of car parks.

The two types of frequency reserves considered, FCR and aFRR, must satisfy the market requirements. For FCR this implies that the reserves need to be available for four consecutive hours and for aFRR the capacity needs to be available only 15 min. Both reserves have a minimum capacity that can be offered equal to 1 MW. While upward and downward aFRR are offered separately, FCR is a symmetrical product. If 1 MW is offered it implies that the power source must be able to increase and decrease the power output with 1 MW. Considering the actual EMS operation of the FCEV, the vehicles cannot start consuming electricity if the CPPP needs to decrease its power output. Thus, to offer FCR and aFRR downwards, the CPPP must offer also electricity.

In order to obtain the occupation pattern of the city car park, data from 7 car parks was collected and averaged. The data set contained the empty parking places

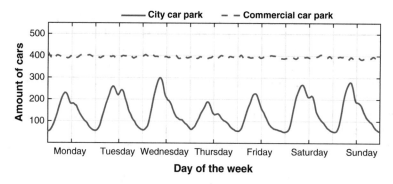

Fig. 8.6 Number of cars that are parked at the car parks during a week for two occupation patterns

measured every 15 min, which is equal to the ISP of the electricity market, and is used as the time step of the simulation.

CPPP Participation in the Electricity Market

Due to the variable available capacity of the vehicles, the forward market is not a reasonable option for the CPPP. The day-ahead and the intraday market, referred to as spot markets, offer enough flexibility for the aggregator to offer its capacity. Almost 90% of the electricity that is sold on the spot markets is sold on the day-ahead market and only 10% is sold on the intraday market. The 2016 prices of the day-ahead market were available on the ENTSO-e transparency platform and were used in the model to calculate the revenues obtained from the offered electricity.

8.2.2.2 Constraints of the System

The total power output of the CPPP was limited by the minimum condition, when all cars were idling, and by the maximum condition ($P_{V2G,max}$), when all cars were offering 10 kW DC. As a consequence also the power that can be offered as electricity (P_e) on the day-ahead market or the reserves offered on the balancing market (P_{FCR}, P_{aFRR}) must remain between those limits, as shown in Eq. (8.4).

$$P_{idling} < P_e, P_{FCR}, P_{aFRR} < P_{V2G,max} \tag{8.4}$$

Based on the amount of electricity that was offered on the day-ahead market, the maximum amount of FCR that could be offered was calculated according to Eq. 8.5.

$$P_{FCR,max} = min((P_e - 0), (P_{V2G,max-P_e})) \tag{8.5}$$

This calculation guarantees the symmetrical characteristics of this particular frequency reserve. However, this amount needed to be available for four consecutive hours. The minimum calculated value in four consecutive hours was assumed as the power reserved for FCR. This value represented the maximum amount of FCR that could be offered by the CPPP. Therefore also 50 and 0% of this value was calculated as an option.

For all the possible combinations of offered electricity and FCR, the amount of aFRR up and down was calculated, as shown respectively in Eqs. (8.6) and (8.7). This was done for all the possible combinations of electricity and FCR offered.

$$P_{aFRR,up} = P_{V2G,max} - (P_e + P_{FCR}) \qquad (8.6)$$

$$P_{aFRR,down} = P_e - P_{FCR} - P_{idling} \qquad (8.7)$$

The available power in the specific ISP was only considered active if the price for the electricity or the reserves was higher than the marginal costs of the CPPP. To evaluate the effect of the position of the CPPP on the bid ladder a best and worst case marginal cost was calculated based on the hydrogen price and the efficiency of the system.

If the power was active, the amount of energy produced for each product (electricity, FCR, aFRR) was calculated. The amount of electricity offered was the maximum amount it could produce in an ISP. For FCR an approximation was made based on historical data of the frequency in continental Europe. The amount of energy used for aFRR was also based on 2016 historical data of the activated reserves.

8.2.2.3   Revenues and Costs

The total revenues per year (R_{CPPP}) were calculated as the sum of the revenues obtained from the three different products, as shown in Eq. (8.8).

$$R_{CPPP} = R_{e,year} + R_{FCR,year} + R_{aFRR,year} \qquad (8.8)$$

The yearly profit for each product is equal to the summation over all the ISPs (N). The amount of required energy per product, in kWh, is multiplied with the respective price, in euro/kWh, for each ISP to obtain the revenues.

The yearly costs (C_{CPPP}) are equal to the operational costs (C_{OC}) of the CPPP summed over the ISP periods in a year, as shown in Eq. (8.9). The latter consist of the cost of purchased energy (C_{pe}), which in this case is the cost of the hydrogen required, and the degradation costs (C_d) the vehicles.

$$C_{CPPP} = \sum_{1}^{N} C_{OC,ISP} = \sum_{1}^{N} (C_{pe,ISP} + C_{d,ISP}) \qquad (8.9)$$

Equation 8.10 shows how the cost of the purchased energy was calculated:

$$C_{pe} = m_{H_2} \times p_{H_2} \tag{8.10}$$

where p_{H_2} is the price of hydrogen per kilogram and m_{H_2} is the total amount of hydrogen required. The price for hydrogen at a fueling station is expected to be circa 2 to 4 dollars per kilogram around mid-century, which corresponds to 1.76–3.52 euro per kilogram with an exchange rate of 0.88 euro per dollar [40–43]. Considering a more hydrogen oriented economy and the assumption of a national hydrogen network, it should be possible to buy hydrogen on an industrial scale from the network. The current price for industrially reformed hydrogen from natural gas is around 1 euro per kilogram [44–46]. This value was used for the financially most optimistic case. To calculate the variable costs related to the hydrogen consumption the range of 1.00–3.52 euro per kilogram H_2 was utilized. These values represent, respectively, the best and worst cases considering the hydrogen price and will be used in further sensitivity analysis.

The total mass of hydrogen used by the CPPP for one year was calculated as the sum of the amount of hydrogen consumed for each ISP for all energy products delivered, as shown in Eq. (8.11).

$$m_{H_2} = \sum_{1}^{N} m_{H_2,ISP} \tag{8.11}$$

Equation (8.12) shows how the mass of hydrogen for every time step was calculated in the model,

$$m_{H_2,ISP} = \frac{E_{V2G,ISP}}{HHV_{H_2} \times \eta_{CPPP,ISP}} \tag{8.12}$$

where $E_{V2G,ISP}$ is the electricity delivered to the grid, $\eta_{CPPP,ISP}$ is the efficiency of the system and HHV_{H_2} is the higher heating value of hydrogen equal to 39.41 kWh/kg. The efficiency is variable and depends on the power output of the vehicles. To estimate the efficiency of the total CPPP system, an average power output per FCEV ($P_{FCEV,ISP}$) in V2G mode was calculated by dividing the total power output of the CPPP (P_{V2G}) by the amount of vehicles that were available in a specific ISP. In this way we assume that all vehicles provide the same amount of power. The efficiency of the total CPPP was assumed equal to the Tank-To-Grid efficiency (TTG) of one FCEV. It represents the efficiency that includes the transformation from hydrogen to the AC electricity that is delivered to the grid. It was calculated with Eq. (8.13), which is the resultant fit from experiments performed with a FCEV in V2G delivering power in the range of 0–10 kW DC [47].

$$\eta_{CPPP,ISP} = \eta_{TTG} = \frac{47 \times P_{FCEV,ISP}}{0.7 + P_{FCEV,ISP}} \tag{8.13}$$

It is expected that the efficiency of the CPPP will increase in the next years because the Tank-to-Wheel efficiency is also expected to increase [26]. The expectation is that the Tank-To-Wheel (TTW) efficiency will increase from 51.5% near future to 61.0% around mid-century. All input variables of this model are based on expected values after 2030. For this reason the CPPP efficiency was increased linearly with the increase of the TTW efficiency. This implies that the used CPPP efficiency in the model was circa 10% higher respect to the one obtained in Eq. (8.13).

The costs for degradation were calculated as extra use of the FCEV due to V2G running time of the fuel cell system. It was assumed that all the electricity was produced by the FC. The degradation of the battery was not taken into consideration. Equation (8.14) shows how the yearly costs for degradation were calculated,

$$C_d = \frac{C_{FC}}{L_h} \times E_{V2G} \times 0.5 \qquad (8.14)$$

where, C_{FC} is the cost per kWh (equal to the capital costs per kW of the fuel cell system), L_h is the lifetime in hours, E_{V2G} is the total amount of electricity produced per year, and 0.5 is a factor to correct for uneven degradation because of driving and V2G. In previous research it was assumed that every produced kWh for electricity balancing was causing 50% of the degradation as produced kWh in driving mode [26]. Therefore, the factor of 0.5 is used in the model. As well, using the cumulative produced energy, instead of power or voltage loss as degradation indicator for dynamic operated fuel cells, is in line with other research approaches [48]. For mid-century it is expected that the lifetime of the FC in driving condition will be around 8000 h. The investment costs of a fuel cell system including maintenance are expected to be circa 26.9 euro per kW around mid-century [26].

Marginal Costs of the CPPP

Table 8.1 presents the used values to calculate the marginal costs. The bid price of a power plant was assumed equal to the marginal costs. In the CPPP the total marginal costs were assumed as the sum of the costs of the required hydrogen and the degradation costs to produce one kWh of electricity. During the analysis with the financial model two cases for the marginal costs of the CPPP were considered, which represented a worst case and a best case scenario based on the expected hydrogen price and the efficiency of the CPPP system.

Table 8.1 Overview of the values used to calculate total marginal cost of the CPPP system in the best and worst case scenarios

	H$_2$ price (€/kg)	Efficiency (%)	Degradation costs (€/kg)	Total marginal costs (€/kg)
Worst case	3.5	40	0.0017	0.2260
Best case	1.0	50	0.0017	0.0524

8.2.2.4 Net Present Value and Payback Time

To calculate the NPV and the payback time, the investment costs were estimated. As it can be seen in Eq. (8.15), the total investment costs (C_{ic}) were equal to the sum of the capital costs (CC_i) of each additional component (n).

$$C_{ic} = \sum_{1}^{n} CC_i \qquad (8.15)$$

The payback time (t_{PB}), was calculated as shown in Eq. (8.16),

$$t_{PB} = \frac{C_{ic}}{CF_{years}} \qquad (8.16)$$

where C_{ic} are the investment costs and CF_{year} is the yearly cash flow. The latter is the difference between the revenues and the costs and is assumed constant each year. The payback period does not take into consideration the time value of the cash flow and may not present the true picture when it comes to evaluating cash flows of a project. Therefore it is combined with the NPV, which considers the discount rate. To determine the NPV, Eq. (8.17) was used:

$$NPV = \sum_{1}^{L} \frac{CF}{(1+r)^t} \qquad (8.17)$$

where CF is the cash flow of each year, t, L is the lifetime of the CPPP and is the discount rate assumed to be 3%. At time t_0 the investment is done, afterwards the yearly cash flow is assumed constant. The NPV is calculated for the worst and best case marginal costs and for different expected lifetimes of the project.

The investment costs represent the adaptations that need to be done to an existing car park to be able to operate as a CPPP with hydrogen as a main power source. This implies that a hydrogen distribution network and a DC grid need to be added to the car park. Also a data communication system needs to be added but the costs of this application are expected to be negligible compared to the other two additions.

Assumptions on the Investment Costs

Every vehicle must be connected with a flexible hose to the local hydrogen network. A vehicle with low-pressure hydrogen inlet for V2G operation has already been proven [49]. Estimating the costs of this system around mid-century is not straightforward as it depends on many factors like the development of the required technologies and the development of the market. Average costs of the nozzle, hose and breakaway ensemble are not publicly available. For this reason we assumed that, due to the characteristic of the transportation and distribution of hydrogen as a gas, it is similar to the transportation and distribution of natural gas in urban areas. This applied as well for the costs of low/pressure hydrogen distributions systems. The costs for the car park were translated into costs per parking place. The costs per parking place included all the costs related to the local hydrogen distribution system. The costs per parking place were assumed equal to the connection costs with the local gas network for a standard household. This is approximately 900 euro per connection [50, 51]. The DC grid represents the ensemble of the wires and all the converters (DC/DC and AC/DC) required to connect the vehicles to the DC grid and the CPPP to the national electricity grid. Also for this system it is difficult to estimate what the costs will be around mid-century as low voltage DC equipment is currently not yet available on a large scale. There are also no standards yet, regarding low voltage DC networks which makes it difficult to estimate the exact settings of the required components. An increase in the DC low voltage market, as it is expected to occur in mid-century, will stimulate the development of the technology and cause a reduction in costs. The costs to connect the FCEVs to the national electricity grid are estimated to be 1750 euros, equal to current costs for a V2G connection point in a car park. We assumed that these costs included all the components required to operate the FCEV in V2G mode. The calculation of the total investment costs is a rough estimation based on the current costs of similar systems. The sum of the costs for the local hydrogen distribution system and the DC grid are equal to 2650 euro per connected parking place. This gives a total of 1,113,000 euro for 420 parking places. Due to the high uncertainty of this value, it is taken into consideration in the sensitivity analysis.

8.2.2.5   Sensitivity Analysis

The uncertainties introduced by the model assumptions and the values of the inputs were assessed through a sensitivity analysis. The NPV and payback time values were calculated modifying the input parameters in a range of $\pm 10\%$. The input parameters that were adjusted were the hydrogen price, lifetime of the fuel cell system, investment costs per vehicle, day-ahead price, aFRR up price, aFRR down price, FCR price and the maximum amount of connected vehicles in the car park. The sensitivity analysis was performed only for the city car park. To calculate the NPV a lifetime of 15 years was assumed. For both marginal costs scenarios the analysis was performed.

8.3   Results and Discussion

8.3.1   Experimental Analysis of the Reaction Time of the FCEV in V2G

Figure 8.7 shows the DC combined power of the FC and battery during the tests with two different V2G power outputs, 5 and 10 kW DC namely. The sum of the power values does not correspond exactly at all times to the V2G power output. At times, the

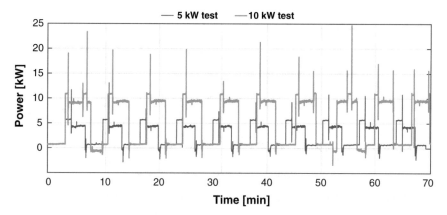

Fig. 8.7 Total DC power supplied by both fuel cell and battery in the FCEV during the dynamic tests

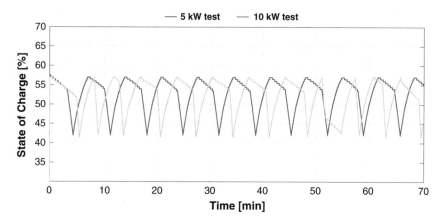

Fig. 8.8 State of charge of the high voltage battery in the FCEV during both tests

sum of battery and FC power is higher than the V2G input signal and other times the value even becomes negative. The reason for this behavior is the Energy Management System (EMS) of the FCEV. The EMS is a crucial part of all EVs. It is the control system, which manages all flows of energy in the car. Every vehicle manufacturer determines different settings in the EMS, which strongly influence the performance of the vehicles. In general the strategy objective in hybrid vehicles, which has a fuel and a battery, is to minimize the fuel consumption. There is, however, a difference between the EMS for driving and V2G mode. Within the Hyundai ix35, when it operates in V2G, the EMS appears to be mainly driven by the state of charge (SoC) of the battery [27]. During both experiments the SoC of the battery always stayed between 42 and 57%, as shown in Fig. 8.8.

To maintain the SoC of the battery within this range, the fuel cell is used to charge the battery. This causes the individual power output of the fuel cell to be higher than

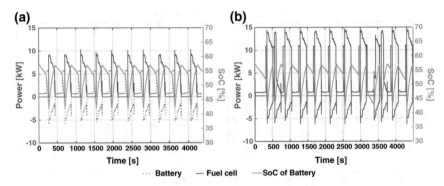

Fig. 8.9 Power flows for fuel cell and battery during the **a** 5 kW and **b** 10 kW DC V2G tests (left y-axis). The State of Charge of the battery is indicated in the right y-axis

the power asked to be delivered to the grid, as can be seen in Fig. 8.9, where the power values for each component are plotted. Negative power values of the battery correspond to periods that the battery was being charged. This is correlated to an increase of the SoC. From Fig. 8.9, the main EMS operations were deduced. When the FCEV was delivering 5 kW DC there was only one mode of operation, which was repeated all 10 times and resulted as follows. During idling, when there was no power requested from the grid, the battery was the main power source that was serving the BoP components. As soon as the car started delivering power to the grid, the battery was providing the power until the SoC reached its lower limit. Then the fuel cell was activated to charge the battery and provide power to the grid as well. Even though the V2G power output was switched off, so the car was not providing power to the grid anymore, the fuel cell was kept on until the battery was charged up 57% SoC. Then the battery remained powering the BoP in idling and this procedure was repeated all test cycles. We refer to this mode of operation as "B+FC", where B stands for battery and FC for fuel cell. On the contrary, there was not a single mode of operation for the ten tests performed when delivering 10 kW DC to the grid, but four different ones. The previously described mode "B+FC" was the most frequent, which occurred in cycles nr. 2, 3, 4, 5, 6, 9 and 10 of the 10 tests performed. In cycle 1, battery was activated first, then FC, then battery again, which was fully drained and at the end the fuel cell was activated one more time ("B+FC+B+FC" mode). In cycle 7 a "B+FC+B" mode was employed to deliver the V2G power. Finally, cycle 8 was the only one that started with the fuel cell delivering power, since the battery was at its lowest SoC. This cycle was "FC+B+FC". Since the power demand is much higher in these tests at 10 kW DC, the alternating behavior between battery and fuel cell to deliver power to the grid is increased and thus results in different modes of operation for each test. They highly depend on the initial state of the battery, because the EMS prioritizes the use of the battery in the SoC range previously mentioned. This mechanism is meant to extend the lifetime of the battery as no deep (dis)charge cycles occur, which is detrimental to the battery's life.

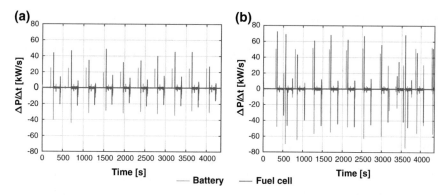

Fig. 8.10 Power gradient versus time for the **a** 5 kW DC tests and the **b** 10 kW DC tests

Table 8.2 Absolute (kW/s) and relative (%/s) upward and downward power gradient values for the fuel cell and battery during the V2G tests

Test	Fuel cell$_{up}$	Battery$_{up}$	Fuel cell$_{down}$	Battery$_{down}$
5 kW	38.3 kW/s (38.3%)	24.7 kW/s (102.8%)	−21.7 kW/s (−21.7%)	−29.6 kW/s (−123.3%)
10 kW	60.2 kW/s (60.2%)	50.0 kW/s (208.2%)	−44.4 kW/s (−44.4%)	−57.2 kW/s (−238.6%)

8.3.1.1 Power Gradient

The power gradient of the fuel cell and the battery measured for the 5 kW and 10 kW DC experiments, are shown respectively in Fig. 8.10a and b. Comparing the results of the two experiments, for higher V2G power settings, the absolute power gradient is higher.

The average of the peak values are shown in Table 8.2 as absolute value and as relative values to the maximum power output of the individual components (100 kW for FC and 24 kW for battery).

It can be observed that for both tests, the absolute power gradient of the fuel cell upward ($FuelCell_{up}$) is higher than the absolute power gradient of the battery upward ($Battery_{up}$). This is not expected since PEMFCs have a lower transient than batteries due to the fact that it is faster to get electricity from stored electrons than drawing hydrogen from the tank, combining it with oxygen and producing the equivalent electricity in the PEMFC [52]. But this observed effect is due to the uneven sizing of the components. That is why we also calculated the relative power gradient. It can be seen that the relative power gradient of the battery is higher for up- and downward regulation than for the fuel cell. For both components, the 10 kW DC experiment had almost a double power gradient compared to the 5 kW DC setting. The current Hyundai ix35 FCEV examined exceeds the performance of a flywheel, which is also considered a suitable device for fast response [53]. For example a 100

Table 8.3 Activation time of the fuel cell and battery in up- and downward reaction in the FCEV when operating in V2G

Test (kW)	Fuel cell$_{up}$ (s)	Battery$_{up}$ (s)	Fuel cell$_{down}$ (s)	Battery$_{down}$ (s)
5	0.13	0.20	0.23	0.17
10	0.17	0.20	0.23	0.17

kW flywheel reaches a full range response in 4 s [54]. This implies that the relative power gradient is 25%/s. This is in most cases a lower gradient than the battery and the FC of the Hyundai ix35. Only when the fuel cell ramps down with a 5 kW DC V2G setting, it presented a lower gradient which was equal to 21.7%/s. This would suggest that for faster response times, the battery would always have to react first and then the fuel cell can provide power for extended periods of time at any of the two power settings.

8.3.1.2 Full Activation Time (FAT)

As can be seen from Table 8.3, the high power gradients result in a sub second reaction time (t_r) of the independent components of the FCEV and thus also for the total V2G power output. The upward reaction time is in the range of 0.13–0.20 s and the downward reaction time results in the range of 0.17–0.23 s. In order to estimate the FAT, the previously obtained reaction time has to be summed to the reaction delay, as was established in Eq. (8.1). A pilot in The Netherlands (NewMotion) was consulted over the reaction delay with aggregated EVs. They offer frequency reserves by increasing or decreasing the charging speed of BEVs. This can be done adapting the software in the charging pole or in the EMS of the vehicle. The reaction delay of these pilots varies strongly depending on the type of aggregation and the used communication system. According to the estimations of the participating pilots, the delay is between 2 and 7 s. The actual maximum allowed reaction delay for FCR is, however, only 2 s. The ability to offer FCR of the aggregated EVs is thus limited due to the relatively long reaction delay. For aFRR the estimated reaction delay is allowed. Since for the FCEV, the reaction time is almost negligible in comparison with the reaction delay, the latter is the one that defines the FAT. The FCEV is certainly able to offer the frequency reserves but the right communication system needs to be adapted to make sure that the reaction delay is small enough.

8.3.2 Financial Analysis

Figure 8.11 presents the results of the yearly cash flow for the different car parks and different power settings offered per car in the car park. The commercial car park has higher yearly cash flows for all cases analyzed than the city car park. This is given

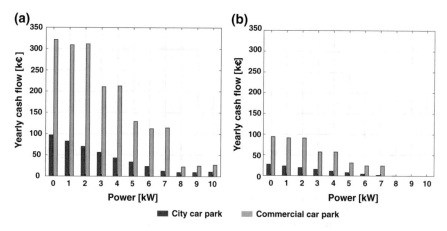

Fig. 8.11 Yearly cash flow results with **a** best and **b** worst case marginal costs for the two types of occupation patterns and considering different values of power for electricity production

Table 8.4 Results of the financial analysis for the two car parks analyzed. The lower limit represents the worst case scenario and higher limit the best case scenario

Type of car park	Yearly cash flow	NPV	Payback time
City	(27–97) k€	(−787–45) k€	(40.8–11.5) years
Commercial	(94–322) k€	(12–2,726) k€	(11.8–3.5) years

the high and constant occupancy of the commercial car park compared to the city car park, where more power capacity from the vehicles is available. In general, when all cars provide higher power, less aFRR can be provided, and this result in lower yearly cash flows. But when lower power is used, competitiveness of the CPPP on the spot market increases, resulting in higher yearly cash flows. The maximum yearly cash flow is obtained when the CPPP offers no electricity and only aFRR upward. As this option is the most profitable, this configuration is assumed for the calculation of the NPV and the payback time. The yearly cash flow, NPV and payback time of the investment of the aggregator to operate as a CPPP system for the two types of occupation patterns are shown in Table 8.4. All of these economic indicators are calculated for the worst and best case for the marginal costs and are expressed in the table as the minimum and maximum limits of the value range, respectively.

The range between worst and best case for the results obtained is very high, indicating how sensible the system is to the marginal costs. The main reason is the lower competitiveness of the car park on the bid ladder for the worst case in comparison with the best case. In general the commercial car park presented a more favorable business case than the city carpark. It can be seen in Table 8.4 that the commercial car park has a payback time that is four times shorter compared to the city car park. The main reason is that there were more moments during the day that the vehicles were parked and could offer reserves. This resulted in higher yearly cash

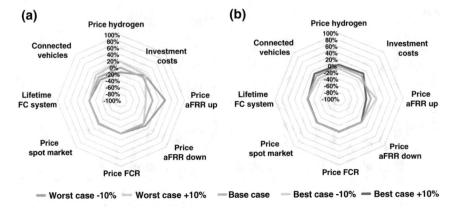

Fig. 8.12 Spider charts showing the results from the sensitivity analysis of the influence of the in put parameters on the payback time for the **a** worst and **b** best case scenarios

flows, and thus shorter payback time. The most interesting investment would result for a commercial carpark with approximately 400 cars available every day. In the best case scenario the investment would be recovered in 3.5 years, obtaining a yearly cash flow of €322,000 and a NPV of €2,726,000.

8.3.2.1 Sensitivity Analysis

Results presented in Sect. 8.3.2 were obtained with a fixed set of assumptions. Additional analysis was conducted to assess how changes in the main input parameters impact the payback time and NPV of the CPPP system. Figure 8.12 shows the effect of varying the input parameters on the payback time for the worst case (Fig. 8.12a) and the best case marginal costs (Fig. 8.12b). The 0% line indicates the outcome with the original values of the parameters. It can be seen that the effect of the price for aFRR down, the price for FCR, the price on the spot market and the lifetime of the FC system on the payback time, is nihil in both cases. The average deviation of the parameters that influence the most the outcome is approximately 10%. These parameters are the investment costs, the amount of vehicles connected, the price for hydrogen and the price for aFRR up.

The amount of connected vehicles is a variable that can be influenced by the aggregator. This is an interesting parameter to adjust to have an influence on the financial potential of the CPPP. A higher occupation, as was seen from the results with the commercial car park, would increase the moments that the aggregator can satisfy the minimum bid size requirement. For city car parks, different incentives could be thought of to favour constant occupation of the car park. This could be done by stimulating users to park longer times by reduced parking tariffs for example. The other three input variables that caused variation in the outputs of the model, were the investment costs, the hydrogen price and the price for aFRR up. The same behavior

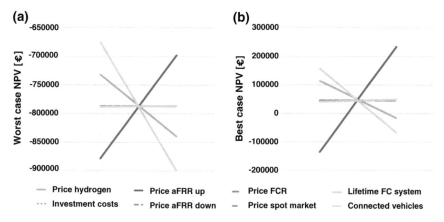

Fig. 8.13 Results from the sensitivity analysis of the effect of the main input parameters on the NPV of the financial analysis

was seen when considering the NPV of the worst and best case for the marginal costs. Results are shown, respectively, in Fig. 8.13a and b (the investment costs data falls exactly under the connected vehicles data). The effect of the changes increased significantly when considering the NPV of the best case for marginal costs. While for all other situations the changes in the outcome were on average ±10%, for the best case marginal costs the changes were factor 10 higher. This effect is caused by the fact that the NPV was calculated over a period of 15 years. If the yearly cashflow increases or decreases with 10%, this effect is added to the NPV each year. The occupation pattern of the car park has a large influence on the payback time and the NPV of the aggregator. When considering a constant occupation pattern in the car park, the cash flow of the CPPP could be increased significantly which results in a shorter payback time and higher NPV.

8.4 Conclusions

This research has shown the technical capability and potential income generation that aggregated Vehicle-To-Grid (V2G) services provided with Fuel Cell Electric Vehicles (FCEVs) could have, with focus on the Netherlands. The reaction time of a fuel cell electric vehicle able to provide power to the grid was experimentally tested. An economic analysis was performed for an existing car park owner, acting as aggregator to provide energy services as *Car Park as Power Plant* (CPPP). For this purpose, an in-house developed financial model was used to evaluate the NPV and payback time of a car park acting as aggregator and offering frequency reserves.

The experiments show that both power sources of the fuel cell electric vehicle, which are the fuel cell stack and the battery, are suitable to offer fast frequency reserves. For the vehicle tested, the ramp up reaction times were in the range of 0.13 - 0.20 s and for ramping down 0.17 - 0.23 s. Besides taking into account the reaction time of the FCEVS, the full activation time (FAT) takes into account the reaction delay in the communication part with aggregator and transmission system operators. The latter was not measured in this research and was obtained from other pilot projects in The Netherlands with battery electric vehicles offering frequency reserves. Both terms together result in values around 2 and 7 s, being the reaction delay the one that defines the FAT. This value is enough to offer automatic frequency restoration reserves (aFRR) but not frequency containment reserves (FCR), since the limit for the latter is 2 s. We recommend for future research/pilot projects to test and improve the communication methods between vehicles and external parties. In this way, the CPPP would be able to increase its frequency reserves portfolio.

This study offered an exploration of the potential economic value that aggregated FCEVs could generate by providing frequency reserves with their fuel cells onboard. The results from the model show that when only the fuel cell stack is used as power source in V2G, the most financial interesting option is to offer only aFRR upwards. Providing this service can yield monetary benefits if the car park has a high and constant occupation. With a car park of approximately 400 cars all year long available, payback times of 11.8 and 3.5 years were obtained taking into account worst and best case scenarios for a 15 year period analysis, respectively. The most sensitive parameters that affect the model results were the amount of vehicles, the price for hydrogen and the price for aFRR up. Future work should be directed at using both power sources in the FCEV, since then there would be more moments that the CPPP could also offer other types of frequency reserves. This would be translated in a higher financial return for the aggregator party.

Acknowledgements C. B. Robledo and A.J.M. van Wijk would like to acknowledge the CESEPS project, which has received funding from the EU Horizon 2020 research and innovation program under the ERA-Net Smart Grids plus grant agreement No 646039, from the NWO and from BMVIT/BMWFW under the Energy der Zukunft program. This work was also financially supported by the Netherlands Organisation for Scientific Research (NWO) [Program "Uncertainty Reduction in Smart Energy Systems (URSES)", Project number 408-13-001] and GasTerra B.V.

References

1. M. Scherer, Frequency control in the European power system considering the organisational structure and division of responsibilities. Ph.D. thesis (2016). https://doi.org/10.3929/ethz-a-010692129Rights
2. P.S. Kundur, N.J. Balu, M.G. Lauby, *Power System Stability And Control*. EPRI Power System Engineering Series (McGraw-Hill, 1994), https://books.google.nl/books?id=v3RxH_GkwmsC
3. M. Huber, D. Dimkova, T. Hamacher, Energy **69**, 236 (2014), http://dx.doi.org/10.1016/j.energy.2014.02.109
4. M.d.l.T. Rodríguez, M. Scherer, D. Whitley, F. Reyer, *IEEE PES General Meeting* (2014). https://doi.org/10.1109/PESGM.2014.6939825

5. W. Kempton, S. Letendre, Transp. Res. Part D: Transp. Environ. **2**(3), 157 (1997). 1361-9209/97 https://doi.org/10.1016/S1361-9209(97)00001-1
6. A. van Wijk, L. Verhoef, Our car as power plant (2014). https://doi.org/10.3233/978-1-61499-377-3-i, http://www.medra.org/servlet/aliasResolver?alias=iospressISBN&isbn=978-1-61499-376-6&spage=7
7. W. Kempton, J. Tomić, J. Power Sour. **144**(1), 280 (2005). https://doi.org/10.1016/j.jpowsour.2004.12.022
8. P. Codani, M. Petit, SSRN (2014). https://doi.org/10.2139/ssrn.2525290
9. M.R. Sarker, Y. Dvorkin, M.A. Ortega-Vazquez, IEEE Trans, Power Syst. **31**(5), 3506 (2016). https://doi.org/10.1109/TPWRS.2015.2496551
10. W. Kempton, J. Tomic, S. Letendre, A. Brooks, T. Lipman, Vehicle-to-grid power: battery, hybrid, and fuel cell vehicles as resources for distributed electric power in California. Technical report, California Air Resources Board and the California Environmental Protection Agency (2001)
11. X. Zhang, S.H. Chan, H.K. Ho, S.C. Tan, M. Li, G. Li, J. Li, Z. Feng, Int. J. Hydr. Energy **40**(21), 6866 (2015). https://doi.org/10.1016/j.ijhydene.2015.03.133
12. IEA, Large-scale electricity interconnection - Technology and prospects for cross-regional networks. Technical report (2016)
13. ENTSO-E, p. 158 (2013)
14. M.J. Poorte, Car park as power plant offering frequency reserves. A technical and economic feasibility assessment. Ph.D thesis, Delft University of Technology (2017)
15. Products, *Day-Ahead Auction* (2018), https://www.epexspot.com/en/product-info/auction
16. Products, *Intraday Continuous* (2018)
17. European Commission, *Reducing CO2 Emissions from Passenger Cars* (2017), https://ec.europa.eu/clima/policies/transport/vehicles/cars_en
18. R.A.C. Van Der Veen, R.A. Hakvoort, *2009 6th International Conference on the European Energy Market, EEM 2009*, pp. 1–6 (2009). https://doi.org/10.1109/EEM.2009.5207168
19. Honda Begins Sales of All-new Clarity Fuel Cell - Clarity Fuel Cell Realizes the World's Top-class Cruising Range Among Zero Emission Vehicles of Approximately 750 km (2016), http://world.honda.com/news/2016/4160310eng.html?from=r
20. Hyundai Motor Company (HMC), *Hyundai ix35 Fuel Cell* (2016), https://www.hyundai.com/worldwide/en/eco/ix35-fuelcell/highlights
21. Hyundai Media Center. NEXO: The Next-Generation Fuel Cell Vehicle From Hyundai (2018), http://www.hyundainews.com/en-us/releases/2456
22. Toyota Motor Corporation (TMC), *Toyota Global Newsroom: Outline of the Mirai* (2014), https://newsroom.toyota.co.jp/en/download/13241306
23. M. Gurz, E. Baltacioglu, Y. Hames, K. Kaya, Int. J. Hydr. Energy **42**(36), 23334 (2017). https://doi.org/10.1016/j.ijhydene.2017.02.124
24. J.K. Kissock, in *Proceedings of the 1998 International Solar Energy Conference* (1998), pp. 121–132
25. T.E. Lipman, J.L. Edwards, D.M. Kammen, Energy Pol. **32**(1), 101 (2004). https://doi.org/10.1016/S0140-6701(04)90146-4, http://linkinghub.elsevier.com/retrieve/pii/S0140670104901464
26. V. Oldenbroek, L.A. Verhoef, A.J.M.V. Wijk, Int. J. Hydr. Energy 1–31 (2017). https://doi.org/10.1016/j.ijhydene.2017.01.155
27. V. Oldenbrock, V. Hamoen, S. Alva, C. Robledo, L. Verhoef, A.V. Wijk, in *6th European PEFC and Electrolyser Forum* (2017), pp. 1–21. 978-3-905592-22-1
28. R. van der Veen, R. Verzijlbergh, Z. Lukszo, A.V. Wijk, in *10th International Conference on Sustainable Energy and Environmental Protection* (2017), pp. 27–30. https://doi.org/10.18690/978-961-286-054-7.15
29. The Green Village (2018), https://www.thegreenvillage.org/about-us
30. T. Lim, B. Ahn, ECS Trans. **50**(2), 3 (2012). https://doi.org/10.1149/05002.0003ecst
31. SAE International, SAE electric vehicle and plug in hybrid electric vehicle conductive charge coupler. Technical report, SAE International (2017), https://saemobilus.sae.org/content/j1772_201710

32. RVO, **2**, 1 (2016), http://www.bovag.nl/data/sitemanagement/media/2013_cijferselektrischvervoertmdecember2013.pdf
33. H. platform, Inbreng H2 Platform voor het Aanvalsplan Duurzame Mobiliteit. Stimulering waterstoftankstations en zero emissie brandstofcel elektrische voertuigen 2018–2022, met doorkijk naar 2030. Technical report (2018)
34. K. Alanne, S. Cao, Renew. Sust. Energy Rev. **1** (2016). https://doi.org/10.1016/j.rser.2016.12.098
35. P.E. Dodds, W. McDowall, Uk Shec (7), 3 (2012)
36. J. Woudstra, P. van Willigenburg, B. Groenewald, H. Stokman, S. De Jonge, S. Willems, in *2013 Proceedings of the 10th Industrial and Commercial Use of Energy Conference* (2013), pp. 2–7, http://ieeexplore.ieee.org/xpls/abs_all.jsp?arnumber=6761675
37. J.J. Justo, F. Mwasilu, J. Lee, J.W. Jung, Renew. Sust. Energy Rev. **24**, 387 (2013). https://doi.org/10.1016/j.rser.2013.03.067, http://dx.doi.org/10.1016/j.rser.2013.03.067
38. A.T. Elsayed, A.A. Mohamed, O.A. Mohammed, Electr. Power Syst. Res. **119**, 407 (2015). https://doi.org/10.1016/j.epsr.2014.10.017
39. P. Van Willigenburg, J. Woudstra, T. De Lange, H. Stokman, *Proceedings of the 22nd Conference on the Domestic Use of Energy, DUE 2014* (2014). https://doi.org/10.1109/DUE.2014.6827758
40. International Energy Agency, Technology roadmap. Hydrogen and fuel cells. Technical report (2015), http://www.springerreference.com/index/doi/10.1007/SpringerReference_7300
41. Tractebel Engineering S.A., Hinicio, **228** (2017), http://www.fch.europa.eu/sites/default/files/P2H_Full_Study_FCHJU.pdf
42. A.J. van Wijk, The green hydrogen economy in the Northern Netherlands. Technical report (2017)
43. J. Eichman, A. Townsend, J. Eichman, A. Townsend (2016)
44. N. Sulaiman, M.A. Hannan, A. Mohamed, E.H. Majlan, W.R. Wan, Daud. Renew. Sust. Energy Rev. **52**, 802 (2015). https://doi.org/10.1016/j.rser.2015.07.132
45. S. Dillich, T. Ramsden, M. Melaina, Hydrogen production cost using low-cost natural gas. Technical report (2012)
46. Energy Renaissance (2018), http://www.h2energyrenaissance.com/
47. C.B. Robledo, V. Oldenbroek, F. Abbruzzese, A.J. van Wijk, Appl. Energy **215**(2017), 615 (2018). https://doi.org/10.1016/j.apenergy.2018.02.038
48. M. Jouin, M. Bressel, S. Morando, R. Gouriveau, D. Hissel, M.C. Péra, N. Zerhouni, S. Jemei, M. Hilairet, B. Ould, Bouamama. Appl. Energy **177**, 87 (2016). https://doi.org/10.1016/j.apenergy.2016.05.076
49. C.B. Robledo, V. Oldenbroek, J. Seiffers, M. Seiffers, A. van Wijk, in *2017 Fuel Cell Seminar and Energy Exposition* (2017), pp. 7–9
50. A. van Wijk, C. Hellinga, Hydrogen - the key to the energy transition. Technical report (2018), http://edepot.wur.nl/333952
51. Liander. Liander. Tarieven 2017 voor consumenten. (2017), https://www.liander.nl/consument/aansluitingen/tarieven2017/?ref=14389
52. J. Wishart, Fuel cells versus batteries in the automotive sector. Technical report (2014)
53. Beacon Power, Frequency regulation compensation in the safe harbor statement. Technical report (2010), https://www.ferc.gov/EventCalendar/Files/20100526085637-Judson, BeaconPower.pdf
54. M. Lazarewicz, *Flywheel Technology Energy Storage for Grid Services Safe Harbor Statement* (2011), https://www.uml.edu/docs/15_Energy_M_Lazarewicz_tcm18-48972.pdf

Part VI
Planning and Scheduling in Renewable-Enriched Energy Grids

Chapter 9
Multi-agent Planning Under Uncertainty for Capacity Management

Frits de Nijs, Mathijs M. de Weerdt and Matthijs T. J. Spaan

Abstract Demand response refers to the concept that power consumption should aim to match supply, instead of supply following demand. It is a key technology to enable the successful transition to an electricity system that incorporates more and more intermittent and uncontrollable renewable energy sources. For instance, loads such as heat pumps or charging of electric vehicles are potentially flexible and could be shifted in time to take advantage of renewable generation. Load shifting is most effective, however, when it is performed in a coordinated fashion to avoid merely shifting the peak instead of flattening it. In this chapter, we discuss multi-agent planning algorithms for capacity management to address this issue. Our methods focus in particular on addressing the challenges that result from the need to plan ahead into the future given uncertainty in supply and demand. We demonstrate that by decoupling the interactions of agents with the constraint, the resulting algorithms are able to compute effective demand response policies for hundreds of agents.

9.1 Introduction

The electric power system is the largest man-made system in the world. Its network is designed such that its connected users and devices all can receive the electrical energy they need whenever they demand. However, the energy system is undergoing a transition. Power is not only generated by controllable power plants, but gradually a larger part comes from intermittent and uncontrollable renewable generation from sun and wind. As the power produced needs to be equal to the power consumed at all times and storage of electrical energy is in many places extremely inefficient, this transition to more renewable generation can only be realized by consuming

F. de Nijs (✉) · M. M. de Weerdt · M. T. J. Spaan
Delft University of Technology, Delft, The Netherlands
e-mail: f.denijs@tudelft.nl

M. M. de Weerdt
e-mail: m.m.deweerdt@tudelft.nl

M. T. J. Spaan
e-mail: m.t.j.spaan@tudelft.nl

© Springer Nature Switzerland AG 2019
P. Palensky et al. (eds.), *Intelligent Integrated Energy Systems*,
https://doi.org/10.1007/978-3-030-00057-8_9

the power at the moment it is produced, called *demand response*. Especially new electrical loads, such as heat pumps and electric vehicles, are relatively flexible: they can be shifted in time to match (renewable) generation.

Shifting loads to moments of high renewable generation creates higher correlations of such controllable loads, and this may create new peak loads in the distribution system. At some places therefore costly reinforcements of the network may seem necessary given the goal of designing for peak use. However, in some areas, such an overload of the network will only occur for a few hours, and only for a few times a year. In such cases, it may not be economical to reinforce the network. The alternative is to coordinate some of these loads to prevent the congestion and guarantee that network use stays within the capacity limits of cables and/or converters. An important difficulty here is that on the one hand this requires looking ahead some time into the future to be able to decide which loads to shift to an earlier (or later) time, but on the other hand we do not know exactly how much renewable power is produced, and how much of the network capacity will be used by uncontrollable loads. This chapter discusses scalable algorithmic methods for capacity management that can deal with such uncertainty.

In the next section, we provide a formal model of the computational problem of capacity management. We give sufficient detail such that it could be straightforwardly implemented to be solved by mixed-integer linear solvers. However, this straightforward approach has two shortcomings. First, it scales poorly with the number of controllable loads, so solving this problem takes too long to be of practical use. Second, it does not capture the uncertainty appropriately (e.g. of generation and of the demand from controllable and uncontrollable loads). In the remaining part of this chapter we introduce methods to get around these shortcomings: we use Markov decision processes (MDPs) to include uncertainty explicitly, and describe how to decouple the problem into agents that only interact through the capacity constraints.

9.2 Problem Description

As a running example in this chapter we use the scheduling of heat pumps. The power draw of a heat pump device easily exceeds the entire remaining household demand. Simultaneous use by a large number of heat pumps easily overloads the capacity of the local grid. However, a heat pump has also a significant potential to contribute to both the integration of renewable sources as well as for capacity management by exploiting the available system inertia: for well-insulated buildings, running the heat pump a few hours earlier can obtain the same level of comfort at negligible extra energy loss. The problem we study then is to optimize the temperature trajectory in the buildings by controlling such thermal devices over time, subject to the available capacity of the electricity network.

In principle, given a discrete-time model of the devices in the aggregation, the control problem can be solved using standard centralized optimization techniques. We present a mathematical formulation of such a general optimization framework for heat pumps here, which is representative of the control problem for any flexible load. Controlling an aggregation of thermal devices subject to a network capacity constraint comes down to choosing an activation schedule per device which ensures the capacities are never exceeded, while simultaneously guaranteeing that every device maintains its desired temperature. This problem can be formulated as a constrained optimization problem, where the temperature goal (the comfort level) is the objective and the capacity is enforced as one of the constraints.

Let $\theta_{i,t}$ be the temperature of a single device i at a specific point in time t, and let $m_{i,t}$ represent the binary (on/off) control decision of the heat-pump. Then, the given temperature transition model f_i specifies how the temperature of the device evolves up to the next decision step $t + 1$ to $\theta_{i,t+1}$. In the following part, we assume that all devices can be modeled by the same general transition model f, with device specific parameters a_i. In the following sections we use the thermal model given by Mortensen and Haggerty [1], which is straightforward to optimize due to its linearity, however our algorithms can be applied to more advanced building thermal models such as those described in [2].

The n heat-pump devices should be constrained such that the sum of power draw does not exceed the (remaining) network capacity and power production. To state the multi-device model we use boldface characters to represent vectors of device parameters over all devices, i.e., $\boldsymbol{\theta}_t = \begin{bmatrix} \theta_{1,t} & \theta_{2,t} & \dots & \theta_{n,t} \end{bmatrix}$. Then, using the Hadamard product $\boldsymbol{b} = \boldsymbol{a} \circ \boldsymbol{\theta}_t \implies \forall i: b_i = a_i \times \theta_{i,t}$, we can define a state transition function to compute $\boldsymbol{\theta}_{t+1}$ as

$$\boldsymbol{\theta}_{t+1} = f(\boldsymbol{\theta}_t, \boldsymbol{m}_t, \theta_t^{\text{out}}) = \boldsymbol{a} \circ \boldsymbol{\theta}_t + (1 - \boldsymbol{a}) \circ \left(\theta_t^{\text{out}} + \boldsymbol{m}_t \circ \boldsymbol{\theta}^{\text{pwr}} \right). \quad (9.1)$$

With this function, we can define a planning problem using a given horizon h, the thermal properties of the n thermostatic loads with initial temperatures $\boldsymbol{\theta}_1$, the predicted outdoor temperature θ_t^{out}, and the predicted power constraint L_t. A solution to such a problem is a device activation schedule that never switches on more devices than is allowed while minimizing cost function $c(\boldsymbol{\theta}_t)$. The entire planning problem becomes:

$$\begin{aligned}
& \underset{[\boldsymbol{m}_1, \boldsymbol{m}_2, \dots, \boldsymbol{m}_h]}{\text{minimize}} \sum_{t=1}^{h} c(\boldsymbol{\theta}_t) \\
& \text{subject to} \quad \boldsymbol{\theta}_{t+1} = f(\boldsymbol{\theta}_t, \boldsymbol{m}_t, \theta_t^{\text{out}}) \\
& \qquad\qquad \sum_{i=1}^{n} m_{i,t} \le L_t \\
& \qquad\qquad m_{i,t} \in [0, 1] \qquad\qquad \forall i, t
\end{aligned} \quad (9.2)$$

Due to the generality of the model, we can optimize the devices for different objectives expressed through the cost function. Besides typical functions such as the squared error on the deviation from the set-point $c(\boldsymbol{\theta}_t) = \sum_{i=1}^{n} \left(\theta_{i,t} - \theta_{i,t}^{\text{set}} \right)^2$, or the maximum deviation $c(\boldsymbol{\theta}_t) = \max_i \left(\theta_{i,t} - \theta_{i,t}^{\text{set}} \right)$, we might imagine more

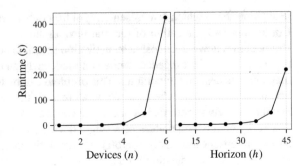

Fig. 9.1 Wall-clock time needed to optimize the mixed-integer linear program (9.2) as a function of the number of devices and the planning horizon

application-specific functions. For example, a refrigerator may only incur high penalties when the temperature gets above a (thawing) threshold.

Unfortunately, this approach suffers from two major drawbacks: limited scalability, and difficulty representing uncertainty. Figure 9.1 demonstrates that solving such a straightforward, centralized model quickly becomes intractable. Furthermore, in the problem description thus far, we ignored that the effect of actions of the heat pump on the (modeled) state of the household is not known exactly, but may vary significantly, because the model does not capture all aspects of reality. For example, the physical building is much more complex, consisting of several different rooms and corridors and windows and walls, radiation from the sun can increase temperatures but is not taken into account, and inhabitants may open and close doors and windows.

In the next section, we demonstrate how to overcome both these weaknesses. First, we show how uncertainty can be incorporated in a principled manner by transforming the proposed model to a multi-agent Markov decision process. Then, we introduce several algorithms to coordinate the agents' demand in a scalable manner, while optimizing their cost functions.

9.3 Multi-agent Planning Under Uncertainty

Here we first introduce an MDP model for the problem of planning the use of heat pumps under uncertainty in the temperature development of the households over time. We show that this can be neatly modeled as a so-called multi-agent MDP (MMDP). In the type of MMDPs that we consider, we identify several agents that take actions more or less independently. In this case, these agents represent the heat pumps of the different houses.

The challenge here is that the size of a straightforward MMDP model increases exponentially with the number of agents, because the set of actions in such a model is the set of *all possible combinations* of actions by individual agents. The domain of study, though, has some specific structure: the only interaction between the agents is because of the limited network capacity. In the remainder of the section we then discuss a number of different methods to exploit this structure by decoupling the reasoning for each of the agents.

First, we show how an arbitrage mechanism can be used in combination with several iterations of optimal responses by the agents given their likelihood of getting allocated to ensure that these individual policies never cross the resource limit.

Second, we investigate preallocation methods to avoid dependence on an on-line component, allowing policies to be executed without communication. Although scalable methods to compute preallocations only satisfy the constraints in expectation, we demonstrate that constraint violations can be minimized by reducing the available constraint capacity. Further, we show that this can even be extended to settings where these constraints themselves are uncertain. For example, it may be that there is an uncertain amount of uncontrollable use of the network, making the exact amount of remaining capacity uncertain as well.

Finally, we discuss a method for the case where the agents do not necessarily have a good model of the building they are controlling.

9.3.1 Centralized MMDP Model

MDPs form a flexible mathematical framework for optimizing the course of action of an agent that experiences an uncertain response from the environment to its actions [3]. This allows it to cope with uncertain environmental factors such as outdoor temperature but also uncertainty resulting from imperfect models, for instance a lack of information about whether windows are closed or not. As our problem consists of multiple decision makers, we model the problem as a Multi-agent Markov decision process (MMDP) [4]. A key assumption is that agents are fully cooperative, i.e., that they optimize their decision making according to a joint objective function.

In our MMDP model for the optimization problem defined in Eq. 9.2 we have a set of n agents that all have the same actions $\mathcal{A} = \{\text{off}, \text{on}\}$ available to them, and an agent-specific state $\mathcal{S}$. The continuous temperature is discretized into k non-overlapping states s_j each defining a temperature interval $[\theta_{s_j,\min}, \theta_{s_j,\max})$. In addition to this, there are two extrema states $s_{\min}$ and $s_{\max}$ ranging from $(-\infty, \theta_{s_1,\min})$ and $[\theta_{s_k,\max}, \infty)$ respectively, resulting in the following state space of an agent: $\mathcal{S} = \{s_{\min}, s_1, s_2, \ldots, s_k, s_{\max}\}$.

The transition function $T : \mathcal{S}^n \times \mathcal{A}^n \times \mathcal{S}^n \to [0, 1]$ describes for each joint state and joint action pair $(\mathbf{s}, \mathbf{t})$ the probability of attaining joint state $\mathbf{s}'$. It is derived by applying the Markov heat-transfer function $f(\theta, m) = \mathbf{a}\theta + (1 - \mathbf{a})(\theta_{\text{outside}} + m\theta_{\text{heating}})$ to the lower and upper values of the temperature range $[\theta_{s,\min}, \theta_{s,\max})$. This produces a new range $[\theta'_{\min}, \theta'_{\max})$ that may overlap the ranges of multiple discrete states $s_j, s_{j+1}, \ldots$. The degree of overlap determines the (uniform) probability of transitioning to each of these potential future states.

Agents are rewarded for their actions in a certain joint state, through the reward function $R : \mathcal{S}^n \times \mathcal{A}^n \to \mathbb{R}$. The rewards assigned to the agents in each time step are the costs depending on how large the deviation from the setpoint $\theta_{i,t}^{\text{set}}$ is:

$$\sum_{i=1}^{n} -\max\{0, |s_i - \theta_{i,t}^{set}| - 0.5\}^2. \tag{9.3}$$

The imposed power constraint L_t which limits the number of activated devices is then encoded in the joint transition function. The joint transition function specifies the cross product of all agents' action spaces $\mathcal{A}^n$. By removing those actions where the number of agents 'on' is more than L_t we obtain the required constraint.

The resulting MMDP can be solved optimally [3], but suffers from scaling exponentially in the number of agents.

9.3.2 Decoupling and Best Response with Arbitrage

A first approach to avoid the exponential blow-up of the centralized model is to model the decision problem for each heat pump as a separate problem, and have a very simple centralized *arbitrage* mechanism to guarantee that never too many devices switch on. The details of this approach can be found in [5]. In this section, we summarize the main conceptual idea behind this method, and discuss the effects on the run time compared to the centralized approaches introduced above.

The problem for each agent representing a heat pump is the MMDP for one heat pump from the previous section, discarding the restriction L_t in the joint transition function. Such so-called *single-agent* MDPs can be solved for each agent separately.

However, to prevent that the agents' plans violate the power limit in a certain time step t, the agent that expects to lose the least utility from switching off is switched off at t, and its plan is re-computed. This arbitrage mechanism is repeated until the conflict is resolved. To determine which agent expects to lose the least utility by going from on to off we look at the difference between the planned utility scores in the value table. However, because this procedure risks getting caught in a local minimum, we use the utility loss as a probability of being selected instead of always selecting the agent that expects to lose the least utility. Moreover, we explicitly model the probability of being forced to switch off by the arbitrage mechanism in the planning model for the single-agent MDPs. This indirectly models the effect of the plans of other agents. We call this approach the arbitrage best-response method (arbitrage-BR).

We use the following artificial instances of the problem to compare the scalability of this approach relative to the optimal solutions. In its simplest form, this instance has 3 agents, a horizon of 20 and $\theta_{i,t}^{set} = 20$, $\forall i, t$. In the first 5 time steps, 3 agents are allowed to switch on, in the next 5 time steps only 2 are allowed, followed by 5 time steps where only 1 is allowed. The final 5 time steps are unconstrained. We then evaluate the scalability by varying the number of agents between 1 and 6, and the horizon between 5 and 45, and the number of agents fixed at 3 (10 instances per setting). In addition, we set the MMDP approach to use only 6 temperature states, and we imposed a run-time cut-off of 5 minutes. The arbitrage-BR method scales

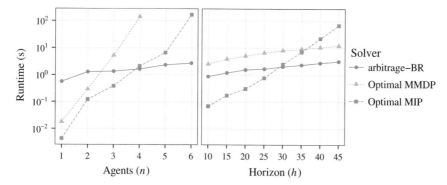

Fig. 9.2 The runtime of both optimal methods does not scale with the number of agents, while the arbitrage-BR method scales reasonably well in both agents and horizon

well in terms of both number of agents and horizon, as shown in the experimental results in Fig. 9.2.

Next, to better understand the potential of arbitrage-BR in practice, we consider a more realistic evaluation in a neighborhood of 182 households equipped with heat pumps. Given runtimes observed earlier, we cannot expect to be able to run the optimal solvers for 182 households. We, therefore, consider a relaxation of the optimal MIP that allows the devices to be switched on only partially. Because in practice these devices cannot be switched on in arbitrary fractions, and because the decision time step granularity of 1 minute is the minimum to prevent short-cycling, it is not possible to implement the outcome of the relaxation. However, this provides a lower bound on the penalty of the optimal solution with binary activations.

For this evaluation, we model a two-day period during which a gradual decrease of the available power occurs starting from hour 6. At hour 20 the minimum capacity is reached and only 10 out of 182 heat pumps are allowed to be switched on. At the start of the second day, all devices can be switched on again. The households would all like to maintain an internal temperature of $21°$ ($\theta_{i,t}^{\text{set}} = 21, \forall i, t$).

The decision frequency is set to once every minute. While it is unlikely that in a real-world scenario the power limit is known with such accuracy, this granularity allows each unit to switch just in time, and it also serves as a worst-case problem size to demonstrate scalability. Since the agents are now decoupled, we can solve the MMDP with much finer temperature discretization. For arbitrage-BR we discretized temperature from 16 to $24°$ over 80 states, resulting in bins of $0.1°$ width.

Figure 9.3 presents the average indoor temperature, the normalized cumulative error and the number of devices switched on for this instance.

First, we observe that all approaches using planning perform significantly better than the non-anticipatory control, which has a rather high cumulative penalty. The cumulative penalty of the arbitrage-BR stays close to the MIP relaxation lower bound, which confirms that it is close to optimal even in this larger instance. Computing the 182 policies for the adaptive decomposition took only 7.5 min, less time than it took

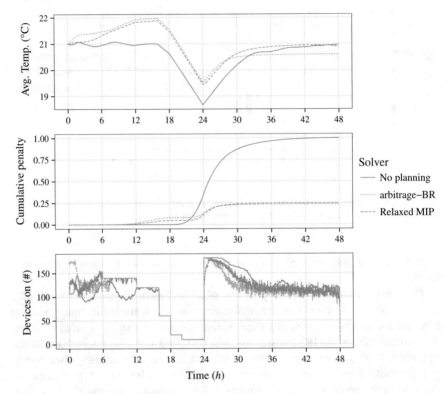

Fig. 9.3 Simulation of the response of a realistic neighborhood of 182 households to a strong curtailment request. Algorithm arbitrage-BR performs on par with the theoretical upper bound given by the relaxed MIP solution

the optimal MMDP solver to compute a solution for the four agent toy example above. This demonstrates that the adaptive decomposition is indeed scalable to real-world instances.

9.3.3 Off-Line Control Through Preallocations

The algorithms discussed in the previous section allow us to safely control an aggregation of heat-pumps through communication with a centralized arbiter. However, because the power grid should be robust to both system failures and malicious attacks, we additionally require that a decentralized fall-back exists. Therefore, in this section, we study algorithms to compute decentralized control policies which satisfy the constraints by adhering to a resource *preallocation*.

A preallocation specifies for each agent in advance at which times it has permission to use resources. Policies computed for a preallocation are communication-free: because the allocation fully specifies the way the constraint should be shared, an agent never needs to coordinate its consumption with others during execution. Given the imposed power constraints per time step L_t, a preallocation $U_{i,t}$ is computed for each agent i, such that the allocations jointly satisfy

$$\forall t: \sum_{i=1}^{n} U_{i,t} \leq L_t. \tag{9.4}$$

Then, each agent can individually optimize a policy π_i satisfying its own allocation, meaning that the consumption of the policy $C_{\pi_i,t}$ never exceeds the preallocation,

$$\forall i: \max_{\pi_i} V_{\pi_i}, \text{ subject to } \forall t: C_{\pi_i,t} \leq U_{i,t}. \tag{9.5}$$

Existing algorithms to compute preallocations for MDPs can be categorized according to the type of preallocation they compute: (i) A MILP [6] and LDD + GAPS [7] compute preallocations which restrict the *worst-case* resource consumption. (ii) The Constrained MDP LP (CMDP; [8]) and Column Generation [9] compute preallocations which restrict the *expected* resource consumption. Unfortunately, both categories have drawbacks which limit their use in practice. The algorithms which restrict worst-case consumption have exponential worst-case complexity in the number of resources, which makes them intractable for our models. In addition, they may lead to low efficiency: resources may sit unused when the uncertain state trajectory leads agents to a state where resources are not needed (e.g. sufficiently warm in the case of heat-pumps). Restricting the expected consumption is tractable, however, the resulting policies are stochastic and may *violate* the constraints at execution time.

Because the tractability of restricting the expected consumption makes it more promising, we investigate how to limit the risk of policies jointly violating the constraints in the next Sect. 9.3.3.1. Then, because renewable power sources may introduce uncertainty about the constraint itself, an extension to compute preallocations for stochastic constraints is proposed in the following Sect. 9.3.3.2.

9.3.3.1   Bounding Constraint Violation Risk

Stochastic preallocation algorithms like CMDPs and Column Generation compute stochastic policies that only ensure that their *expected* resource consumption does not violate the limits. As such, these methods do not provide any guarantees regarding the probability that a resource limit is violated during execution. In this section we introduce methods which improve these algorithms by ensuring that the probability of violating any individual constraint is upper bounded by a given parameter α. In doing so, we summarize our work on bounding the probability of constraint violations, for details see [10].

Because of the uncertainty present in the transition function of the temperature, the temperature state $s_{i,t}$ of agent i in a future time step t is a random variable. Given a control policy π_i which switches the heat-pump on below a certain temperature, the future power consumption of an agent also becomes a random variable $C_{i,\pi_i,t}$. When each agent executes their policy unconditionally, and without communication, these random variables are independent. Therefore, by independence, we can compute the total resource consumption at time t as the sum of the agents' consumption, $C_t = \sum_{i=1}^{n} C_{i,\pi_i,t}$. The stochastic allocation algorithms guarantee that

$$\forall t : \mathbb{E}[C_t] \leq L_t. \tag{9.6}$$

In practical applications, even when constraints are soft, exceeding them typically incurs some cost to the system operator, such increased wear from overheating when exceeding the capacity of a power grid element. Therefore, even when we are using stochastic allocation algorithms, we would additionally like to restrict the probability that a realization of C_t exceeds the limit, or

$$\forall t : \mathbb{P}[C_t > L_t] \leq \alpha. \tag{9.7}$$

To obtain policies which additionally satisfy the constraint on the tail probability, we propose to impose reduced resource constraints $0 \leq L_t^* \leq L_t$, resulting in more conservative policies. Because the random variables $C_{i,\pi_i,t}$ can be upper bounded by the power consumption of the most-consuming action, we can apply Hoeffding's inequality [11] to determine L_t^*, resulting in

$$L_t^* = L_t - \sqrt{\frac{\ln(\alpha) \cdot \left(\sum_{i=1}^{n} (\max C_{i,t})^2\right)}{-2}}. \tag{9.8}$$

In practice the bound obtained by applying Hoeffding's inequality can be relatively loose (Fig. 9.4, top). Therefore, we also propose a dynamic constraint relaxation technique which adjusts the reduced resource limit L_t^* on the basis of empirical evidence of actual violations during simulation (Fig. 9.4, bottom).

To evaluate the proposed approaches to bound the risk of constraint violations, we compare them to the heat-pump planning problem. Each agent has its possible temperature states discretized over 24 states, and we plan for a time horizon of 24 steps. Figure 9.5 presents the performance of the algorithms as the number of agents grows. We observe that the preallocation algorithms constraining worst-case performance (MILP and LDD + GAPS) indeed exhibit poor scalability, as they exceed 60 minutes of computation time at 4 agents. At the same time, we observe that while the CMDP algorithm is highly scalable, it computes solutions which exceed the available capacity nearly half the time on tight constraints. The results show that this high risk is averted when we virtually reduce the resource capacity available to the planner through application of Hoeffding's inequality. However, the resulting policies are on the conservative side of the risk threshold, staying an order of magnitude below the target tolerance of $\alpha = 0.05$. Our dynamic constraint relaxation algorithm is able to target the lower, observed tolerance of $\alpha = 0.005$ exactly. While

Fig. 9.4 Histograms showing realized resource demands obtained through simulation. Policies to satisfy the constraint (*solid lines*) are computed for reduced limits (*dashed lines*).
Top: reduced resource limits on the basis of Hoeffding's inequality.
Bottom: initial and final iterations of our dynamic bound relaxation

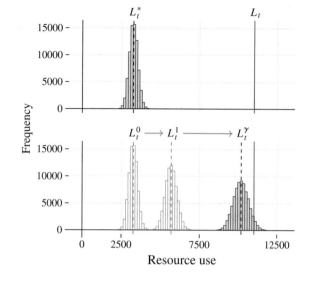

Fig. 9.5 Comparison of the performance of preallocation algorithms on heat-pump planning problems as the number of agents increases, measured on three performance metrics.
Top: expected value of the solution normalized to CMDP (log x).
Middle: simulated constraint violation probability of the most-often violated constraint (log-log).
Bottom: mean wall-clock time required to compute a policy, in minutes (log x)

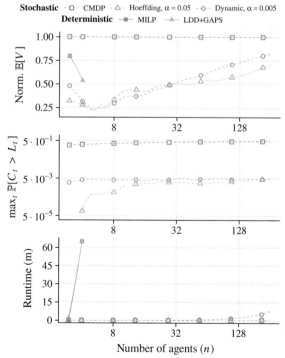

the dynamic algorithm takes slightly longer to compute, it nevertheless remains tractable, especially when compared to the deterministic allocation algorithms. In addition, the solutions it finds are of higher quality than the Hoeffding-bounded policies for larger numbers of agents, even though the risk level it attains is the same. In conclusion, when comparing stochastic preallocations we observe that for large numbers of agents, the values of the bounded approaches tend towards the CMDP value. When more agents are available to spread the load of the reduced resource limit, their individual rewards are compromised less.

9.3.3.2 Computing Preallocations for Stochastic Constraints

Thus far we have assumed that the planner knows exactly how much power is available in each time step. One major challenge of the integration of renewable energy sources such as wind and solar power generators is that it makes the available power production capacity dependent on the weather, and therefore volatile. Unfortunately, it is not yet possible to predict weather perfectly even on short (day-ahead) time-scales. Therefore controllers of such buffers should take into account multiple statistical forecast scenarios [12, 13]. In order to address this requirement, this section investigates how this assumption can be relaxed to deal with stochastic resource constraints when communication is unreliable; for more details, see our work in [14].

A collection of potential power production scenarios can be represented by a Markov chain defined over outcomes. This Markov chain is defined by a state space S_L of power production outcomes, and the transition probabilities $T_L : S_L \times S_L \rightarrow [0, 1]$. Since all agents must adhere to the same constraint, the transition function of the stochastic constraint threatens to couple the agents together. Fortunately, Becker et al. [15] show that independence is retained when shared features only exist in a part of the state space that agents cannot affect themselves. As such, the stochastic constraint problem can also be *decomposed* into n single-agent sub-problems, which we propose to do by augmenting the state space of each agent with the current limit (captured in factored state space $S_L \times S$). Nevertheless, the preallocation algorithms must be modified to handle the fact that agents expected consumption is now correlated with the probability of visiting a power production state s_L.

Alternatively, we could collapse the constraint Markov chain to its expectation in each time step, to obtain a planning problem with a fixed constraint, which the preallocation algorithms can solve directly. However, doing so would result in policies which make two-sided errors: if the realized constraint is less than the expectation, the policy is likely to cause a violation, while if the realized constraint is more than the expectation, the policy will leave resources unused. In addition, knowledge of the current constraint-state may inform which future scenarios are more likely, allowing the planner to anticipate on future constraint realizations. Therefore, we expect that taking into account the model of stochasticity in the preallocation will result in policies which are both significantly safer and which obtain significantly better expected value.

Fig. 9.6 Comparison of two approaches to handle probabilistic forecasts of power production: using the expected value (*squares*) versus using all scenarios in the planning problem (*triangles*). Results for increasing planning horizon on three performance metrics (lower values are better). **Top**: deviation from the optimal expected value, normalized to horizon length (log-log). **Middle**: simulated number of constraint violations, normalized to horizon length (log x). **Bottom**: mean wall-clock time required to compute a policy, in minutes (log x)

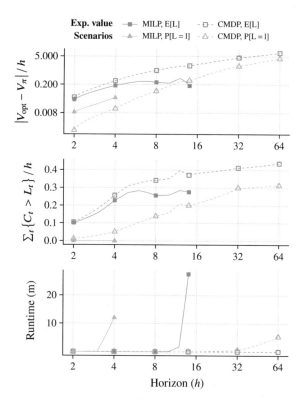

To test our expectations, we perform an experiment comparing the modified pre-allocation algorithms on an instance of the heat-pump planning problem with 10 artificially generated power production scenarios. For these experiments, we discretized the temperature state range into 25 states, and we restrict the number of agents to 3 in order to be able to compute the optimal (on-line) joint policy. This allows us to compare the solution quality of the preallocation algorithms objectively. Figure 9.6 presents the results of an experiment evaluating plan quality as the length of the planning horizon increases. We observe that planning for the stochastic constraint scenarios is more computationally intensive, resulting in increased plan runtime. However, in return, we observe that both our expectations on plan quality hold in practice: planning for a stochastic constraint results in a smaller error *and* fewer constraint violations for both types of preallocation algorithms.

Another conclusion we can draw from this experiment is that the quality of control degrades when agents must operate without communication for long periods of time. While off-line control may be required to satisfy grid robustness requirements, under normal operation we expect agents to be able to operate with regular communication intervals. To determine if there are also benefits to incorporating stochastic constraints in such a rolling horizon re-planning setting, we perform an additional experiment where we let the agents communicate their current state at set intervals. Figure 9.7 presents the results, showing the relationship between the average number

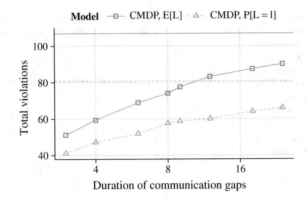

Fig. 9.7 Effect of intermittent communication on the quality of control when: using the expected value of the constraint (*squares*) versus using all constraint scenarios in the planning problem (*triangles*). Plot shows the total number of violations relative to the no-communication upper bound (*horizontal lines*), as the time between successive re-planning increases (log x)

of violations over the horizon (of $h = 216$) and the re-planning frequency, with a gap of three indicating that agents communicate and re-plan every fourth time-step. We observe that re-planning more frequently leads to fewer constraint violations, although this comes at the cost of needing sufficient computational capacity to compute a plan before the next decision point. Further, we see that in this re-planning setting it also makes sense to use the scenario information in the planning problem, as this also significantly reduces the total number of violations. Note that, although we did not do so here for comparison, we can combine stochastic constraints with the bounding technique presented in Sect. 9.3.3.1 to further reduce the risk of constraint violations.

9.3.4   Learning Agent Types

Up to now, we have assumed that all model parameters are fully known to the planning algorithm, in which case a solution can be computed offline (i.e., before execution of the policies). In practice, however, that might not be the case, which requires the planning algorithm to take into account information it can gather online (i.e., while executing the policies). In particular, we consider a setting in which model parameters such as grid constraints are known, but certain characteristics of the individual agents are not. By observing their behavior, however, we can estimate their model parameters. The key challenge is how to optimize the heating of houses given uncertain estimates of their parameters while capacity limits are not to be exceeded.

We identify different *types* of agents, where each type identifies certain key parameters of the system the agent is controlling. Those can be physical characteristics, such as the insulation level of a house, but can also be related to user preferences, such as the desired temperature setpoint. While types in principle can be dynamic, our focus has been on static types [16]. For instance, the physical properties of a building are unlikely to change quickly, although we likely do not know the thermal response of every building initially; to address this challenge requires the use of a learning agent [17]. However, in this setting, we only need to perform the learning once, as part of the initialization of the device.

To deal with the initially unknown type of each agent, we proposed two novel algorithms [16]: the first algorithm is an extension of Posterior Sampling Reinforcement Learning [18] to the multi-agent, constrained setting. The second algorithm exploits the structural properties of the problem to approximately solve the constrained partially observable problem itself, by bounding the belief space expansion to states where the regret of switching to the best type's MDP policy is low. In particular, we showed how both algorithms can be used as subroutine in the Column Generation stochastic preallocation algorithm described above.

9.4 Conclusions

This chapter discusses how to keep our energy system affordable by making demand responsive to grid limitations, shifting some of the demand to less congested times. In order to perform effective demand response, the controller is required to optimize over future control decisions: in order to decide if we can shift charging an electric vehicle to a later time, we need to know when the car must be charged, and how many more charging opportunities will come. Unfortunately, optimizing a control policy over the future necessarily involves dealing with uncertainty, both in the needs of the device under control (e.g. when the owner of the car returns), as well as in the evolution of the system (e.g. the demand of uncontrolled loads, and the production from renewable sources). We show in Sect. 9.3.1 that optimizing control under uncertainty can be naturally modeled as a Markov decision process. Unfortunately, the resulting constrained, multi-agent Markov decision process model of demand response suffers from intractable scalability in the number of devices. In this chapter we present several novel algorithms to overcome this intractability, as well as innovations that make existing algorithms more effective.

In the first place, we show that the scalability challenge can be overcome if we are able to decouple the control problem of the individual devices from the constraint allocation problem. Section 9.3.2 investigates the use of a centralized resource arbiter to distribute available capacity *on-line*, on the basis of the utility each device expects to receive from its requested allocation. This decouples the agents, as each agent can determine an individual best-response to the probability that the arbiter will award its request. The resulting algorithm is shown to efficiently find solutions with a minimal loss in solution quality. In addition, the arbiter guarantees that the solution satisfies the current system constraints at all times.

Unfortunately, the use of an on-line arbiter requires devices to maintain a connection to the centralized mechanism at all times. In order to compute solutions which are robust to both connection failures and malicious attacks, we investigate resource preallocation algorithms in Sect. 9.3.3. Existing tractable algorithms compute preallocations which satisfy the constraint in expectation, which results in a high risk of constraint violations. While constraint violations can sometimes be absorbed by the inertia of the system, their occurrence should nevertheless be avoided. In Sect. 9.3.3.1 we present an effective approach to bound the probability of constraint violations while retaining the tractability of the preallocation algorithms. Stochastic constraints (as resulting from wind prediction scenarios) are an additional challenge for preallocation algorithms because the agents cannot coordinate on the realized constraint on-line. Nevertheless, we show in Sect. 9.3.3.2 that preallocation algorithms can be modified to incorporate stochastic constraints, resulting in solutions which are both of higher quality and resulting in fewer constraint violations. This result can be combined with occasional re-planning to further improve the coordination.

Finally, we show in Sect. 9.3.4 that our results can be extended to a learning setting, where the devices operate according to one of a set of potential models describing behavior types (for example, insulation levels and preferred set-points). By making use of an optimal learning framework, we are able to identify the correct device model in a minimal number of learning steps.

These results show that our proposed algorithms and extensions are effective at computing high-quality demand-response policies. Nevertheless, there are challenges left to address in future work. Importantly, shifting demand may come at costs for the users. In the current model we assume these costs are known and the algorithms aim to minimize the total costs. However, in many situations, besides total costs, also fairness is an important criterion to take into account. For example, we may not want to have always the same (well-insulated) house be pre-heated throughout the night in order to minimize losses, because the owners will in the end have higher energy costs than without coordination. Related is the issue that the true rewards may not be known to the system, but that the system is informed, e.g. about the desired temperature. In the current proposal, there is no remedy against users who feel that they are left out in the cold and just increase their desired set-point significantly above the real goal in order to increase home temperature. This will probably increase their allocation of the scarce capacity, but at the cost of other users.

Nevertheless, the positive results in simulation make it worthwhile to run a pilot in practice to assess the algorithms performance in resolving bottlenecks in real-world infrastructure. In addition, the methods introduced in this chapter are not just made for heat pumps. They can easily include any other demand that can be modeled as a Markov decision process. In fact, the proposed methods are even more general: we have also applied them to a budget allocation problem in on-line advertising, and in computing capacity-aware recommendations for guiding tourists through crowded cities [16].

Acknowledgements Support of this research by network company Alliander is gratefully acknowledged.

References

1. R.E. Mortensen, K.P. Haggerty, A stochastic computer model for heating and cooling loads. IEEE Trans. Power Syst. **3**(3), 1213–1219 (1988)
2. T.A. Reddy, J.F. Kreider, P.S. Curtiss, A. Rabl, *Heating and Cooling of Buildings*, 3rd edn. (CRC Press, USA, 2017)
3. M.L. Puterman, *Markov Decision Processes-Discrete Stochastic Dynamic Programming* (Wiley, New York, 1994)
4. C. Boutilier, Planning, learning and coordination in multiagent decision processes, in *TARK* (1996), pp. 195–210
5. F. de Nijs, M.T.J. Spaan, M.M. de Weerdt, Best-response planning of thermostatically controlled loads under power constraints, in *Proceedings of the Twenty-Ninth AAAI Conference on Artificial Intelligence* (2015), pp. 615–621
6. J. Wu, E.H. Durfee, Resource-driven mission-phasing techniques for constrained agents in stochastic environments. J. Artif. Intell. Res. **38**, 415–473 (2010)
7. P. Agrawal, P. Varakantham, W. Yeoh, Scalable greedy algorithms for task/resource constrained multi-agent stochastic planning, in *Proceedings of the 25th International Joint Conference on Artificial Intelligence* (2016), pp. 10–16
8. E. Altman, *Constrained Markov Decision Processes*. Stochastic Modeling (Chapman & Hall/CRC, Boca Raton, 1999)
9. K.A. Yost, A.R. Washburn, The LP/POMDP marriage: optimization with imperfect information. Nav. Res. Logist. **47**(8), 607–619 (2000)
10. F. de Nijs, E. Walraven, M.M. de Weerdt, M.T.J. Spaan, Bounding the probability of resource constraint violations in multi-agent MDPs, in *Proceedings of the 31st AAAI Conference on Artificial Intelligence* (2017), pp. 3562–3568
11. W. Hoeffding, Probability inequalities for sums of bounded random variables. J. Am. Stat. Assoc. **58**(301), 13–30 (1963)
12. P. Pinson, H. Madsen, H.A. Nielsen, G. Papaefthymiou, B. Klöckl, From probabilistic forecasts to statistical scenarios of short-term wind power production. Wind Energy **12**(1), 51–62 (2009). https://doi.org/10.1002/we.284
13. A. Staid, J.P. Watson, R.J.B. Wets, D.L. Woodruff, Generating short-term probabilistic wind power scenarios via nonparametric forecast error density estimators. Wind Energy **20**(12), 1911–1925 (2017). https://doi.org/10.1002/we.2129
14. F. de Nijs, M.T.J. Spaan, M.M. de Weerdt, Preallocation and planning under stochastic resource constraints, in *Proceedings of the 32nd AAAI Conference on Artificial Intelligence* (2018), pp. 4662–4669
15. R. Becker, S. Zilberstein, V. Lesser, C.V. Goldman, Solving transition independent decentralized Markov decision processes. J. Artif. Intell. Res. **22**, 423–455 (2004). https://doi.org/10.1145/860575.860583
16. F. de Nijs, G. Theocharous, N. Vlassis, M.M. de Weerdt, M.T.J. Spaan, Capacity-aware sequential recommendations, in *Proceedings of International Conference on Autonomous Agents and Multi Agent Systems* (2018), pp. 416–424
17. F. Ruelens, S. Iacovella, B. Claessens, R. Belmans, Learning agent for a heat-pump thermostat with a set-back strategy using model-free reinforcement learning. Energies **8**(8), 8300–8318 (2015). https://doi.org/10.3390/en8088300
18. M.J.A. Strens, A bayesian framework for reinforcement learning, in *Proceedings of the 17th International Conference on Machine Learning* (2000), pp. 943–950

Chapter 10
Computationally Tractable Reserve Scheduling for AC Power Systems with Wind Power Generation

Vahab Rostampour, Ole Ter Haar and Tamás Keviczky

Abstract This work presents a solution method for a day-ahead stochastic reserve scheduling (RS) problem using an AC optimal power flow (OPF) formulation. Such a problem is known to be non-convex and in general hard to solve. Existing approaches follow either linearized (DC) power flow or iterative approximation of nonlinearities, which may lead to either infeasibility or computational intractability. In this work we present two new ideas to address this problem. We first develop an algorithm to determine the level of reserve requirements using vertex enumeration (VE) on the deviation of wind power scenarios from its forecasted value. We provide a theoretical result on the level of reliability of a solution obtained using VE. Such a solution is then incorporated in OPF-RS problem to determine up- and down-spinning reserves by distributing among generators, and relying on the structure of constraint functions with respect to the uncertain parameters. As a second contribution, we use the sparsity pattern of the power system to reduce computational time complexity. We then provide a novel recovery algorithm to find a feasible solution for the OPF-RS problem from the partial solution which is guaranteed to be rank-one. The IEEE 30 bus system is used to verify our theoretical developments together with a comparison with the DC counterpart using Monte Carlo simulations.

This research was supported by the Uncertainty Reduction in Smart Energy Systems (URSES) research program funded by the Dutch organization for scientific research (NWO) and Shell under the Aquifer Thermal Energy Storage Smart Grids (ATES-SG) project with grant number 408-13-030.

V. Rostampour (✉) · O. Ter Haar · T. Keviczky
Delft Center for Systems and Control, Delft University of Technology, Mekelweg 2, 2628 CD Delft, The Netherlands
e-mail: v.rostampour@tudelft.nl

O. Ter Haar
e-mail: ole.terhaar@gmail.com

T. Keviczky
e-mail: t.keviczky@tudelft.nl

© Springer Nature Switzerland AG 2019
P. Palensky et al. (eds.), *Intelligent Integrated Energy Systems*,
https://doi.org/10.1007/978-3-030-00057-8_10

10.1 Introduction

The reserve scheduling (RS) problem deals with day-ahead scheduling of the reserve power to accommodate possible mismatches between forecasted and actual wind power. Stochastic variants of the RS problem, where violations of operational constraints, e.g. power generations, bus voltage and line flow constraints, are allowed with a small probability to achieve better performance, have received a lot of attention in the past few years, see [1–4] and the references therein. A stochastic RS problem is typically formulated using a lossless DC model based on the assumption of constant voltage magnitudes and small voltage angles, while ignoring the active power losses [5]. These assumptions do not hold in general and may lead to sub-optimality or even infeasibility when implemented on real world systems, especially for networks under a high degree of stress [6].

Using an AC model of the power network enables the stochastic RS formulation to accurately model the effect of large deviations of wind power from its forecasted value, and to offer a-priori suitable reserves such that both active and reactive power, and complex-valued voltage are globally optimal. Due to the non-convexity of the OPF problem, identifying such an optimal operating point of a power system may not be straightforward. In [7], different reformulations and relaxations of the AC OPF problem were presented and their connections were discussed. By means of semidefinite programming (SDP), in [7] a convex relaxation was provided under the existence of a rank-one SDP solution to guarantee the recovery of an optimal solution of the power network.

An RS problem that incorporates an OPF formulation has been introduced in [8] where a chance-constrained OPF problem was formulated. With some modifications, the authors in [8] provided a theoretical guarantee that the OPF-RS problem yields a rank-one feasible solution. Using a heuristic sampling approach, they showed that the resulting optimization problem involves an OPF problem for each wind power profile. Our work here is motivated by [8] to provide some results in a more systematical approach.

While preparing the final version of this work, [9, 10] independently gave an approach to solve an OPF-RS problem in each hour separately, based on the results in [4]. The OPF-RS formulation in [9] is similar to [8] with some modifications, whereas in [10] the formulation is weaker compared to [8], since the condition to distribute reserves among generators is relaxed. Even though the authors in [8] presented a complete day-ahead OPF-RS formulation with up- and down spinning reserves, the results in the aforementioned references are limited either to be heuristic or to a single hourly-based RS with the relaxed conditions. The major barrier of representing an OPF-RS problem as a SDP is the necessity of defining a square SDP matrix variable, which makes the cardinality of scalar variables of the OPF-RS problem quadratic with respect to the number of buses in the power network. This may yield a very large-scale SDP problem for realistic large-scale power networks of interest.

Our work here differs from the aforesaid references in two important aspects. We propose an algorithm to determine a worst-case reserve requirement in each hour by vertex enumeration (VE) of all possible deviations of wind power scenarios from the forecast value. The outcome of VE determines the up- and down-spinning reserves in a probabilistic sense. Using the OPF-RS problem formulation, similarly to [8] with some modifications, we distribute the up- and down-spinning reserves among generators together with the generator dispatch planning for day-ahead schedules. To address the resulting high-dimensional SDP problem, we leverage the sparsity pattern in power networks to break down the large-scale positive-semidefinite (PSD) constraints into small-size constraints, similarly to [11, 12]. We then propose a novel recovery algorithm to obtain a rank-one solution based on the results in [13]. It is important to highlight that this work is based on the same authors' conference paper published in [14] and thesis report in [15].

The layout of this work is as follows. Section 10.2 formulates the RS problem using AC OPF model of power systems by including the uncertain wind power generation, whereas in Sect. 10.3, we provide a computationally tractable reformulation to solve the resulting large-scale SDP in infinite dimensional spaces. Section 10.4 provides a simulation result using IEEE benchmark, whereas Sect. 10.5 concludes this work with remarks and future work.

Notations

$\mathbb{R}$, $\mathbb{R}_+$ denote the set of real and positive real numbers, $\mathbb{S}$, $\mathbb{S}_+$ denote the set of symmetric matrices and positive-semidefinite matrices, respectively. $\mathbb{C}$ denotes the set of complex numbers. Vectors are denoted by lowercase-bold letters $\boldsymbol{a} \in \mathbb{R}^n$, and uppercase letters are reserved for matrices $A \in \mathbb{R}^{n \times n}$. The symbols $A^\top$, A^*, and A^H are used for the transpose, complex conjugate and conjugate transpose of a matrix, respectively. The notations $\underline{a}$ and $\overline{a}$ are used to denote the minimum and maximum allowed values, respectively. The cardinality of a set $\mathcal{A}$ is denoted by $|\mathcal{A}|$.

10.2 Problem Formulation

AC OPF Problem

Consider a power system with a set of buses $\mathcal{N}$, a set of lines $\mathcal{L} \subseteq \mathcal{N} \times \mathcal{N}$ and a set of generator buses $\mathcal{G} \subseteq \mathcal{N}$ such that $|\mathcal{N}| = N_b$ and $|\mathcal{G}| = N_G$. The set of wind power generation buses is denoted by $\mathcal{F} \subseteq \mathcal{N}$ such that $|\mathcal{F}| = N_w$. A set of hours $\mathcal{T}$ forms the scheduling horizon of the hourly-based RS optimization problem and in this work $|\mathcal{T}| = 24$. The vectors $\boldsymbol{p} \in \mathbb{R}^{N_b}$, $\boldsymbol{q} \in \mathbb{R}^{N_b}$ and $\boldsymbol{s} \in \mathbb{C}^{N_b}$ denote real, reactive and apparent power, respectively.

Define the decision variables to be the generator dispatch $\boldsymbol{p}_t^G, \boldsymbol{q}_t^G \in \mathbb{R}^{N_G}$ and the complex bus voltages $\boldsymbol{v}_t \in \mathbb{C}^{N_b}$ for each time step $t \in \mathcal{T}$. Using the rectangular voltage notation: $\boldsymbol{x}_t := [\Re\boldsymbol{v}_t^\top \Im\boldsymbol{v}_t^\top]^\top \in \mathbb{R}^{2N_b}$, we follow [7, Lemma 1] to determine the data-matrices $Y_k, Y_k^*, Y_{lm}, Y_{lm}^*, M_k$. The cost function is the cost of real power generation, expressed as a second order polynomial [16], where the coefficient vectors $\boldsymbol{c}^{\mathrm{qu}}, \boldsymbol{c}^{\mathrm{li}} \in \mathbb{R}_+^{N_G}$ correspond to the quadratic and linear cost coefficients, respectively, and $[c^{qu}]$ represents a diagonal matrix with entries c^{qu}. We now formulate the AC OPF problem by taking into account the effect of wind power generations as follows:

$$\underset{\{x_t, p_t^G, q_t^G\}_{t \in \mathcal{T}}}{\text{minimize}} \quad \sum_{t \in \mathcal{T}} (c^{li})^\top p_t^G + (p_t^G)^\top [c^{qu}] p_t^G \tag{10.1a}$$

subject to:

1. Power generation limits $\forall k \in \mathcal{G}, \forall t \in \mathcal{T}$:

$$\begin{aligned} \underline{p_k^G} \leq p_{k,t}^G \leq \overline{p_k^G}, \\ \underline{q_k^G} \leq q_{k,t}^G \leq \overline{q_k^G}. \end{aligned} \tag{10.1b}$$

2. Power balance at every bus $\forall k \in \mathcal{G}, \forall t \in \mathcal{T}$:

$$\begin{aligned} x_t^\top Y_k x_t = p_{k,t}^G - p_{k,t}^D + p_{k,t}^w, \\ x_t^\top Y_k^* x_t = q_{k,t}^G - q_{k,t}^D. \end{aligned} \tag{10.1c}$$

where $p_t^w := \{p_{k,t}^w\}_{k \in \mathcal{F}}$ is the wind power, and $s_t^D := \{s_{k,t}^D\}_{k \in \mathcal{N}}$ is the demanded power such that $s_{k,t}^D = p_{k,t}^D + q_{k,t}^D$. Note that it is assumed[1] $\mathcal{G} \cap \mathcal{F} = \emptyset$ which means there is no wind power at generator buses.

3. Bus voltage limits $\forall k \in \mathcal{G}, \forall t \in \mathcal{T}$:

$$|\underline{v_k}|^2 \leq x_t^\top M_k x_t \leq |\overline{v_k}|^2. \tag{10.1d}$$

4. Lineflow limits $\forall (l, m) \in \mathcal{L}, \forall t \in \mathcal{T}$:

$$\left(x_t^\top Y_{lm} x_t\right)^2 + \left(x_t^\top Y_{lm}^* x_t\right)^2 \leq |\overline{s_{lm}}|^2,$$

which can be reformulated using the Schur Complement [17] to form a linear matrix inequality constraint, such that the fourth order dependence on the voltage vector is reduced to quadratic terms:

$$\begin{bmatrix} -|\overline{s_{lm}}|^2 & x_t^\top Y_{lm} x_t & x_t^\top Y_{lm}^* x_t \\ x_t^\top Y_{lm} x_t & -1 & 0 \\ x_t^\top Y_{lm}^* x_t & 0 & -1 \end{bmatrix} \preceq 0. \tag{10.1e}$$

5. Reference bus constraint $\forall t \in \mathcal{T}$:

$$x_t^\top E_{\text{ref}} x_t = 0, \tag{10.1f}$$

---

[1] This assumption is considered to streamline the presentation and it is not restrictive for our proposed framework.

where E_{ref} is a diagonal matrix from the standard basis vector $e_{N_b+i_{\text{ref}}}$, and i_{ref} denotes the reference bus.

Remark 10.1 The power balance constraints (10.1c) can be used to reformulate the real and reactive generator dispatch in terms of the voltage vector as follows $\forall k \in \mathcal{N}, \forall t \in \mathcal{T}$:

$$p_{k,t}^G = \mathbf{x}_t^\top Y_k \mathbf{x}_t + p_{k,t}^D - p_{k,t}^w, \tag{10.2a}$$

$$q_{k,t}^G = \mathbf{x}_t^\top Y_k^* \mathbf{x}_t + q_{k,t}^D. \tag{10.2b}$$

Using this reformulation, one can substitute for $p_{k,t}^G$ and $q_{k,t}^G$ in (10.1b) to have $\forall k \in \mathcal{N}, \forall t \in \mathcal{T}$:

$$\underline{p_k^G} \le \mathbf{x}_t^\top Y_k \mathbf{x}_t + p_{k,t}^D - p_{k,t}^w \le \overline{p_k^G}, \tag{10.3a}$$

$$\underline{q_k^G} \le \mathbf{x}_t^\top Y_k^* \mathbf{x}_t + q_{k,t}^D \le \overline{q_k^G}, \tag{10.3b}$$

where the lower and upper generation limits have also been extended to $\mathcal{N}$ using $\underline{p_k^G} = \overline{p_k^G} = 0 \, \forall k \in \{\mathcal{N} \setminus \mathcal{G}\}$.

Remark 10.2 Following Remark 10.1, one can reformulate the cost function (10.1a) using the voltage vector $\mathbf{x}_t$:

$$f_G^x(\mathbf{x}_t, \mathbf{p}_t^w, \mathbf{p}_t^D) := \sum_{k \in \mathcal{G}} c_k^{\text{li}} \left(\mathbf{x}_t^\top Y_k \mathbf{x}_t + p_{k,t}^D - p_{k,t}^w \right) + \tag{10.4}$$
$$c_k^{\text{qu}} \left(\left(\mathbf{x}_t^\top Y_k \mathbf{x}_t + p_{k,t}^D - p_{k,t}^w \right) \right)^2.$$

It is important to note that this function is of order four with respect to $\mathbf{x}$, but it can be also made quadratic.[2] To streamline the presentation, these steps are skipped.

Using $\{\mathbf{x}_t\}_{t \in \mathcal{T}}$, we reformulate the problem (10.1) in a more compact form:

$$\text{OPF}(\{\mathbf{p}_t^w\}) : \begin{cases} \underset{\{\mathbf{x}_t\}_{t \in \mathcal{T}}}{\text{minimize}} \;\; \sum_{t \in \mathcal{T}} f_G^x(\mathbf{x}_t, \mathbf{p}_t^w, \mathbf{p}_t^D) \\ \text{subject to} \;\; (10.1\text{d}), (10.1\text{e}), (10.1\text{f}), (10.3) \end{cases},$$

where the time dependency of $\text{OPF}(\{\mathbf{p}_t^w\}_{t \in \mathcal{T}})$ is dropped for clarity of the notation.

$\text{OPF}(\{\mathbf{p}_t^w\})$ is a quadratically constrained quadratic program (QCQP) in $\{\mathbf{x}_t\}_{t \in \mathcal{T}}$, and a non-convex optimization problem, since the data matrices $Y_k, Y_k^*, Y_{lm}, Y_{lm}^*$ are indefinite [7], which is in fact an NP-hard problem [18] and very hard to solve, in general.

---

[2] The cost function can be made linear with the use of the epigraph notation (see also [17, Chapter 4.1.3]). The resulting inequality constraint can be converted to a LMI using the Schur Complement (see also [17, Chapter A.5.5]), which yields a quadratic function of $\mathbf{x}$.

Convexified AC OPF Problem

Using a semi-definite reformulation (SDR) technique (see [7, 19] and the references therein), one can reformulate OPF($\{p_t^w\}$) as an equivalent problem in a matrix variable $W_t := x_t x_t^\top \in \mathbb{S}^{2N_b}$. W_t represents the operating state of the network, and is therefore called the state matrix. We define $\mathcal{W} \subset \mathbb{S}^{2N_b}$ as the set of feasible operating states, such that $W_t \in \mathcal{W}$, using the following characteristics:

$$
\mathcal{W}(p^w, s^D) := \left\{ W \in \mathbb{S}^{2N_b} \mid \mathrm{Tr}\,(E_{\mathrm{ref}} W) = 0, \right.
$$

$$
p_k^G \leq \mathrm{Tr}\,(Y_k W) + p_k^D - p_k^w \leq \overline{p_k^G}, \forall k \in \mathcal{N},
$$

$$
q_k^G \leq \mathrm{Tr}\,(Y_k^* W) + q_k^D \leq \overline{q_k^G}, \forall k \in \mathcal{N}, \tag{10.5}
$$

$$
|v_k|^2 \leq \mathrm{Tr}\,(M_k W) \leq |\overline{v_k}|^2, \forall k \in \mathcal{N}, \forall (l, m) \in \mathcal{L},
$$

$$
\left. \begin{bmatrix} -|\overline{s_{lm}}|^2 & \mathrm{Tr}\,(Y_{lm} W) & \mathrm{Tr}\,(Y_{lm}^* W) \\ \mathrm{Tr}\,(Y_{lm} W) & -1 & 0 \\ \mathrm{Tr}\,(Y_{lm}^* W) & 0 & -1 \end{bmatrix} \preceq 0 \right\},
$$

where p^w is the wind power, and $s^D = p^D + i q^D$ is the demanded power. Consider now the following formulation as an equivalent optimization problem to OPF($\{p^w\}$):

$$
\underset{\{W_t\}_{t \in \mathcal{T}}}{\text{minimize}} \quad \sum_{t \in \mathcal{T}} f_G(W_t, p_t^w, p_t^D) \tag{10.6a}
$$

$$
\text{subject to} \quad W_t \in \mathcal{W}(p_t^w, s_t^D), \qquad \forall t \in \mathcal{T}, \tag{10.6b}
$$

$$
W_t \succeq 0, \qquad \forall t \in \mathcal{T}, \tag{10.6c}
$$

$$
\mathrm{rank}\,(W_t) = 1, \qquad \forall t \in \mathcal{T}, \tag{10.6d}
$$

where f_G is defined in (10.4), using $W_t = x_t x_t^\top$. Constraints (10.6c) and (10.6d) are introduced to guarantee the exactness of SDR and consequently, OPF($\{p_t^w\}$) and (10.6) to be equivalent.

The optimization problem (10.6) is non-convex, due to the presence of rank-one constraint (10.6d). Removing (10.6d) relaxes the problem to a semi-definite program (SDP). It has been shown in [7] and later in [20] that the rank-one constraint can be dropped without affecting the solution for most power networks. In [8, Proposition 1], the authors showed that when the convex relaxation of the AC OPF problem has solutions with rank at most two, then, forcing any arbitrary selected entry of the diagonal of the matrix W_t to be zero results in a rank-one solution W_t^{opt}. This condition is motivated by the fact that in practice the voltage angle of one of the buses (the reference bus) is often fixed at zero. We denote by C-OPF($\{p_t^w\}$) the convexified version of OPF($\{p_t^w\}$), i.e. Problem (10.6) with the rank-one constraint (10.6d) removed.

OPF-RS Problem Formulation

Consider a power network where a TSO aims to solve a day-ahead AC OPF problem to determine an optimal generator dispatch for the forecasted wind power trajectory such that: (1) the equipments of power system remain safe and (2) the power balance (10.1c) in the power network is achieved. As a novel feature in our proposed formulation C-OPF($\{p_t^w\}$) has a dependency on $\{p_t^w\}_{t\in\mathcal{T}}$, and thus, it solves the OPF problem by taking into account the actual wind power trajectory $\{p_t^w\}_{t\in\mathcal{T}}$. We here define the difference between a generic actual wind power realization and the forecasted wind power, as the mismatch wind power at each time step, e.g. $p_t^m = p_t^w - p_t^{w,f}$. Due to the fact that $\{p_t^m\}_{t\in\mathcal{T}}$ is a random variable, the following technical assumption is necessary in order to proceed to the next steps.

Assumption 1 $\{p_t^m\}_{t\in\mathcal{T}}$ are defined on some probability space $(\mathcal{P}, \mathfrak{B}(\mathcal{P}), \mathbb{P})$, where $\mathfrak{B}(\cdot)$ denotes a Borel σ-algebra, and $\mathbb{P}$ is a probability measure defined over $\mathcal{P}$.

In order to ensure the feasibility and validity of the power network (top TSO priority), we formulate the following problem:

$$\underset{\{W_t^f\}_{t\in\mathcal{T}}}{\text{minimize}} \quad \sum_{t\in\mathcal{T}} f_G(W_t^f, p_t^{w,f}, p_t^D) \tag{10.7a}$$

$$\text{subject to} \quad W_t^f \in \mathcal{W}(p_t^{w,f}, s_t^D), \qquad \forall t \in \mathcal{T}, \tag{10.7b}$$

$$W_t \in \mathcal{W}(p_t^w, s_t^D), \ \forall p_t^m \in \mathcal{P}, \qquad \forall t \in \mathcal{T}, \tag{10.7c}$$

$$W_t^f \succeq 0, \ W_t \succeq 0 \qquad \forall t \in \mathcal{T}, \tag{10.7d}$$

where $\{p_t^{w,f}\}_{t\in\mathcal{T}}$ denotes the forecasted wind power trajectory, $\{p_t^w\}_{t\in\mathcal{T}}$ is a generic wind power trajectory, $\{W_t^f\}_{t\in\mathcal{T}}$ is related to the state of the network in the case of forecasted wind power, and $\{W_t\}_{t\in\mathcal{T}}$ is a generic network state for a generic wind power trajectory. Constraints (10.7b) and (10.7c) ensure feasibility for every network state, while constraints (10.7d) enforce positive semidefiniteness of all network states.

As a second task of the TSO, the power balance of the power network has to be achieved to ensure demand satisfaction even in the presence of uncertain wind power generation. To address this issue, the TSO employs *reserve power scheduling*, using the fact that a mismatch between actual wind power and forecasted wind power can be mitigated by the controllable generators [1]. We can thus express

$$r_{k,t} := p_{k,t}^G - p_{k,t}^{G,f}, \tag{10.8}$$

where $r_t = \{r_{k,t}\}_{k\in\mathcal{G}} \in \mathbb{R}^{N_G}$ denotes the amount of reserve requirement in the power network. Following Remark 10.1, we have:

$$p_{k,t}^G = \text{Tr}\left(Y_k W_t\right) + p_{k,t}^D - p_{k,t}^w,$$

$$p_{k,t}^{G,f} = \text{Tr}\left(Y_k W_t^f\right) + p_{k,t}^D - p_{k,t}^{w,f},$$

and one can substitute them in (10.8) to obtain the reserve power in terms of the network states W_t and W_t^f as follows:

$$
\begin{aligned}
r_{k,t} &:= \mathrm{Tr}\left(Y_k\left(W_t - W_t^f\right)\right) - (p_{k,t}^w - p_{k,t}^{w,f}) \\
&= \mathrm{Tr}\left(Y_k\left(W_t - W_t^f\right)\right) - p_{k,t}^m , \\
&= \mathrm{Tr}\left(Y_k\left(W_t - W_t^f\right)\right) ,
\end{aligned}
\tag{10.9}
$$

where the term $p_{k,t}^m$ is dropped, since it is assumed that there is no wind power at generator buses. The elements of $r_t = \{r_{k,t}\}_{k \in N_G}$ can be positive and negative (the upspinning and downspinning reserve power, respectively) such that they are deployed for a power deficit and surplus to bring balance to the network and satisfy the demanded power [16]. Following the automatic generator regulation (AGR) mechanism [4], we also define two vectors $d_t^{us}, d_t^{ds} \in \mathbb{R}^{N_G}$ to distribute the amount of up- or downspinning reserve power among the available generators for each hour $t \in \mathcal{T}$. To obtain the optimal control strategies for AGR, we consider the following equality constraint $\forall p_t^m \in \mathcal{P}, \forall k \in \mathcal{G}$ and $\forall t \in \mathcal{T}$:

$$
\begin{aligned}
r_{k,t} &= \mathrm{Tr}\left(Y_k\left(W_t - W_t^f\right)\right) \\
&= -d_{k,t}^{us} \min\left(0, \mathbf{1}^\top p_t^m\right) - d_{k,t}^{ds} \max\left(0, \mathbf{1}^\top p_t^m\right) .
\end{aligned}
\tag{10.10}
$$

In order to always negate the mismatch wind power using the reserve power and bring balance to the power network, we enforce the sum of the distribution vectors to be equal to one using the following constraint $\forall t \in \mathcal{T}$:

$$
\mathbf{1}^\top d_t^{us} = 1 , \ \mathbf{1}^\top d_t^{ds} = 1 ,
\tag{10.11}
$$

where $\mathbf{1}$ is a vector of appropriate dimensions with all entries equal to one. Define $r_t^{ds}, r_t^{us} \in \mathbb{R}^{N_G}$ such that $\forall t \in \mathcal{T}$:

$$
-r_t^{ds} \le r_t \le r_t^{us} ,
\tag{10.12a}
$$

$$
0 \le r_t^{us} , \ 0 \le r_t^{ds} ,
\tag{10.12b}
$$

and consider corresponding linear up- and downspinning cost coefficients $c^{us}, c^{ds} \in \mathbb{R}_+^{N_G}$ yielding the total reserve cost:

$$
f_R(r_t^{us}, r_t^{ds}) := (c^{us})^\top r_t^{us} + (c^{ds})^\top r_t^{ds}.
$$

Using $\Xi := \left\{ W_t^f, W_t, d_t^{us}, d_t^{ds}, r_t^{us}, r_t^{ds} \right\}_{t \in \mathcal{T}}$ as the set of decision variables, and combining our previous discussions with the optimization problem (10.7), we are now in the position to formulate the OPF($\{p_t^w\}$) problem with RS in a more compact form:

$$\text{C-OPF-RS}: \begin{cases} \min_{\Xi} \; \sum_{t \in \mathcal{T}} \left(f_G(W_t^f, \boldsymbol{p}_t^{w,f}) + f_R(\boldsymbol{r}_t^{us}, \boldsymbol{r}_t^{ds}) \right) \\ \text{s.t.} \quad (10.7b), (10.7c), (10.7d), (10.10), (10.11), (10.12) \end{cases}.$$

Notice that one needs to substitute $\boldsymbol{r}_t$ in (10.12a) with (10.9).

C-OPF-RS is an uncertain infinite SDP program, due to the unknown and unbounded set $\mathcal{P}$. It is therefore computationally intractable and in general difficult to solve. In the next section, we propose a technique to approximate $\mathcal{P}$ such that it contains the probability mass distribution of $\mathcal{P}$ almost surely with a high level of confidence.

10.3 Tractable Reformulations

In this section, we first present a tractable approach to approximately solve C-OPF-RS, and then, we leverage the sparsity in the problem data to decompose the computationally expensive PSD constraints.

Vertex Enumeration Scheme

The constraint function of C-OPF-RS is a linear function with respect to the uncertainty $\boldsymbol{p}_t^m$, if we exclude (10.10). However, due to the nonlinear operators, max-min, it is not straightforward to reformulate (10.10) as a linear constraint. In fact such operators lead to a hybrid operation, and especially in (10.10), the two terms on the right-hand side cannot be non-zero simultaneously. Following this observation, one can approximate the uncertainty set $\mathcal{P}$ using two sets $\overline{\mathcal{B}}, \underline{\mathcal{B}}$, and reformulate (10.10) as follows $\forall k \in \mathcal{G}, \forall t \in \mathcal{T}$:

$$\begin{aligned} \mathrm{Tr}\left(Y_k \big(W_t - W_t^f \big) \right) &= -d_{k,t}^{ds}(\boldsymbol{1}^\top \boldsymbol{p}_t^m), \; \forall \boldsymbol{p}_t^m \in \overline{\mathcal{B}}, \\ \mathrm{Tr}\left(Y_k \big(W_t - W_t^f \big) \right) &= -d_{k,t}^{us}(\boldsymbol{1}^\top \boldsymbol{p}_t^m), \; \forall \boldsymbol{p}_t^m \in \underline{\mathcal{B}}. \end{aligned} \tag{10.13}$$

Remark 10.3 It is important to notice that all other uncertain constraints in C-OPF-RS have to be satisfied for all $\boldsymbol{p}_t^m \in \overline{\mathcal{B}}$ and $\boldsymbol{p}_t^m \in \underline{\mathcal{B}}$, separately, for all $k \in \mathcal{G}$ and for all $t \in \mathcal{T}$.

Our goal here is to approximate the uncertainty set $\mathcal{P}$ by employing recent results in randomized optimization, the so-called scenario approach [21], to characterize $\overline{\mathcal{B}}, \underline{\mathcal{B}}$ and provide feasibility certificates. A similar technique has been also used in [3, 22, 23] based on [24]. It is now of interest to characterize $\overline{\mathcal{B}}, \underline{\mathcal{B}}$ such that $\overline{\mathcal{B}} \bigcup \underline{\mathcal{B}}$ approximates $\mathcal{P}$. We assume for simplicity that $\overline{\mathcal{B}}$ and $\underline{\mathcal{B}}$ are two axis-aligned hyper-rectangular sets. This is not a restrictive assumption and any convex set could have been chosen instead as described in [22]. We define $\overline{\mathcal{B}} := [\boldsymbol{0}, \overline{\boldsymbol{p}}_t^m]$, and $\underline{\mathcal{B}} := [\underline{\boldsymbol{p}}_t^m, \boldsymbol{0}]$ as two intervals, where the vectors $\overline{\boldsymbol{p}}_t^m \in \mathbb{R}^{N_w}$ and $\underline{\boldsymbol{p}}_t^m \in \mathbb{R}^{N_w}$ define the bounds of hyper-rectangular sets. We now propose Algorithm 6 that aims to determine both sets $\overline{\mathcal{B}}$ and $\underline{\mathcal{B}}$ with minimal volume such that $\overline{\mathcal{B}} \bigcap \underline{\mathcal{B}} = \emptyset$. Consider $\overline{\boldsymbol{p}}^{*m}_t$ and $\underline{\boldsymbol{p}}^{*m}_t$

to be the outcome of Algorithm 6 that determine $\overline{\mathcal{B}}^*$ and $\underline{\mathcal{B}}^*$, respectively. Defining $\mathcal{B}^* = \overline{\mathcal{B}}^* \bigcup \underline{\mathcal{B}}^*$, we next provide the following theorem that establishes a theoretical connection between $\mathcal{B}^*$ and $\mathcal{P}$ by means of the level of approximation.

Theorem 10.1 *Fix* $\varepsilon \in (0, 1)$, $\beta \in (0, 1)$,

$$N_s \geq \frac{2}{\varepsilon} \left(2N_w + \ln \frac{1}{\beta} \right) , \tag{10.14}$$

and construct the set $\mathcal{S} = \{ p_t^{m,1}, p_t^{m,2}, \cdots , p_t^{m,N_s} \}$. *Then,*

$$\mathbb{P}^{N_s} \left[\mathcal{S} \in \mathcal{P}^{N_s} : \mathbb{P}[p_t^m \in \mathcal{P} : p_t^m \notin \mathcal{B}^*] \leq \varepsilon \right] \geq 1 - \beta,$$

where $\mathbb{P}^{N_s}$ *denotes an* N_s*-fold product probability.*

Proof The proof is a direct result of [25, Theorem 1]. $\qquad \square$

The interpretation of Theorem 10.1 is as follows. Given a generic sample $p_t^m \in \mathcal{P}$, the probability of $p_t^m \in \mathcal{B}^*$ is greater than $1 - \varepsilon$ with high confidence level $1 - \beta$.

Algorithm 6 Vertex Enumeration (VE) Algorithm

1: Input: ε, β
2: $N_s \leftarrow \lceil \frac{2}{\varepsilon} \left(2N_w + \ln \frac{1}{\beta} \right) \rceil$
3: Extract $\{ p_t^{m,1}, p_t^{m,2}, \cdots , p_t^{m,N_s} \} \in \mathcal{P}^{N_s}$
4: **for** $t \in \mathcal{T}$ **do**
5: $\quad \mathcal{I} \leftarrow \emptyset, \{ \overline{p^*}_t^m \} \leftarrow \emptyset, \{ \underline{p^*}_t^m \} \leftarrow \emptyset$
6: $\quad$ **for** $k \in \mathcal{F}$ **do**
7: $\quad\quad \mathcal{I} \leftarrow \mathcal{I} \cup \arg\max_i \{ p_{k,t}^{m,i} \}$
8: $\quad\quad \mathcal{I} \leftarrow \mathcal{I} \cup \arg\min_i \{ p_{k,t}^{m,i} \}$
9: $\quad$ **end for**
10: $\quad \mathcal{I} \leftarrow \mathcal{I} \cup \arg\max_i \{ \mathbf{1}^\top p_t^{m,i} \}$
11: $\quad \mathcal{I} \leftarrow \mathcal{I} \cup \arg\min_i \{ \mathbf{1}^\top p_t^{m,i} \}$
12: $\quad$ **for** $i \in \mathcal{I}$ **do**
13: $\quad\quad$ **if** $\mathbf{1}^\top p_t^{m,i} > 0$ **then**
14: $\quad\quad\quad \{ \overline{p^*}_t^m \} \leftarrow \{ \overline{p^*}_t^m \} \cup p_t^{m,i}$
15: $\quad\quad\quad \{ \underline{p^*}_t^m \} \leftarrow \{ \underline{p^*}_t^m \} \cup \mathbf{0}$
16: $\quad\quad$ **else if** $\mathbf{1}^\top p_t^{m,i} < 0$ **then**
17: $\quad\quad\quad \{ \overline{p^*}_t^m \} \leftarrow \{ \overline{p^*}_t^m \} \cup \mathbf{0}$
18: $\quad\quad\quad \{ \underline{p^*}_t^m \} \leftarrow \{ \underline{p^*}_t^m \} \cup p_t^{m,i}$
19: $\quad\quad$ **end if**
20: $\quad$ **end for**
21: **end for**
22: Output: $\{ \overline{p^*}_t^m , \underline{p^*}_t^m \}_{t \in \mathcal{T}}$

A complete description of Algorithm 6 can be found in [15, Section 4.1.2]. Combining the previous discussion with Remark 10.3, the resulting problem is tractable by means of the robust SDP program which is a finite dimensional optimization problem. We consider the generic network state $\{W_t\}_{t \in \mathcal{T}}$ to be the sum of forecasted network state $\{W_t^f\}_{t \in \mathcal{T}}$ and a corrective control action. In this way, it emulates the way the set-points of the AGR unit can be adjusted as a function of the wind power. Due to the fact that the constraint functions of C-OPF-RS, by replacing (10.10) with (10.13), will be convex functions with respect to the uncertainty $p_t^m \in \mathcal{B}^*$, it suffices to enforce the constraints only at the values that correspond to the vertices $\overline{p^*}_t^m$ and $\underline{p^*}_t^m$ of $\mathcal{B}^* = \overline{\mathcal{B}^*} \bigcup \underline{\mathcal{B}^*}$. To solve numerically the resulting robust SDP problem, we use a similar approach as in [3, 22] based on [26] together with enforcing the PSD constraints $\overline{W}_t \succeq 0$ and $\underline{W}_t \succeq 0$ in each hour $t \in \mathcal{T}$, separately, for the worst-case uncertainties $\overline{p^*}_t^m$ and $\underline{p^*}_t^m$, respectively.

It is worth mentioning that compared to the result of [9], the number of samples needed from $\mathcal{P}$ is much lower, since the dimension of the decision variable is much smaller. We formulate a robust variant of the OPF-RS problem that uses far less samples of the uncertainty compared to the approach in [9], whilst having the same theoretical probabilistic guarantees.

Sparsity Pattern Decomposition

SDPs with matrix variables subject to PSD constraints are computationally complex. One can reduce the size of the computationally expensive PSD constraints by selecting certain submatrices of the original matrix variables and only imposing PSD property on those matrices. The solution will be a partially filled matrix, with only those entries filled that correspond to at least one of the submatrices. All other entries will be undetermined. Various algorithms are available for *matrix completion*, the a-posteriori filling of the undetermined entries. In [27], Grone et al. provided the *chordal theorem*, that guarantees the completed matrix will be PSD if and only if specific submatrices are PSD. Consider a symmetric matrix $X \in \mathbb{S}^d$, and let G be a graph with nodes $\{1, \ldots, d\}$. The chordal theorem states that one can reconstruct the PSD Hermitian[3] matrix X using only the entries of X that correspond to the nodes in the maximal cliques[4] of G, if and only if G is a chordal graph.[5] The chordal theorem can thus be used to prove the equivalence between the PSD property of a matrix and the PSD property of its submatrices, thereby reducing the size of the PSD constraints with the overall computational complexity.

Consider a graph over all the buses of the power network such that the edges correspond to the non-zeros in all the data-matrices $Y_k, Y_k^*, Y_{lm}, Y_{lm}^*, M_k$, where the

---

[3]Note that a symmetric matrix is a Hermitian matrix with all its imaginary values equal to zero, i.e. $\mathbb{S} \subset \mathbb{H}$, so the chordal theorem also holds for symmetric matrices.

[4]A *clique* is a subset of nodes that together form a complete graph, i.e. the number of edges between any two nodes in a clique is equal to one. A clique is *maximal* if it is not a subset of any other cliques in the graph [28].

[5]A graph is *chordal* if every cycle of length greater than three has a chord (an edge between non-consecutive vertices in the cycle) [28].

aggregate sparsity pattern can be found. Due to the definition of the nodal admittance matrix, the sparsity pattern is identical to the network topology.

Using a greedy decomposition algorithm [29], we decompose the network in K subsets of buses, corresponding to the maximal cliques of the chordal graph. Denote every clique with $C_k \subset \mathcal{N}$, and collect all cliques in $C = \{C_1, \ldots, C_K\}$, and let N_C be the number of buses in the largest maximal clique: $N_C := \max_k |C_k|$. Every subset C_k induces a submatrix from the original matrix by selecting the columns and rows corresponding to the buses in it. Note that the decomposed problem has K matrix variables of dimension N_C at most.

We now decompose the PSD constraints (10.7d) on every matrix variable in C-OPF-RS $\forall t \in \mathcal{T}$ using the following constraints:

$$W_t^f(C_k, C_k) \succeq 0, \qquad\qquad \forall C_k \in C \qquad\qquad (10.15a)$$

$$\overline{W_t}(C_k, C_k) \succeq 0, \qquad\qquad \forall C_k \in C \qquad\qquad (10.15b)$$

$$\underline{W_t}(C_k, C_k) \succeq 0, \qquad\qquad \forall C_k \in C \qquad\qquad (10.15c)$$

We call the proposed decomposed formulation as CD-OPF-RS. The following proposition is the direct result of [13, Theorem 1].

Proposition 10.1 *The optimal objective value of CD-OPF-RS is equivalent to the optimal objective value of C-OPF-RS.*

The obtained solution using CD-OPF-RS is a partially filled matrix, denoted by $\tilde{W}_t^f$. From this matrix, we wish to reconstruct a PSD matrix which is rank-one and an optimal solution for the proposed original OPF-RS. Although the chordal theorem proves the possibility of completing a PSD matrix, it does not provide a PSD matrix with the desired rank. We therefore aim to develop a matrix recovery algorithm such that the resulting solution is a PSD matrix with rank one.

Inspired by the voltage vector recovery algorithm in [13], we propose a matrix recovery algorithm which is guaranteed to complete a partially filled state matrix to a rank-one PSD matrix. We modify their algorithm for the rectangular voltage notation, and extract a complex voltage vector from the partially filled solution. We then recover the full state matrix from the complex voltage vector. Algorithm 7 summarizes our proposed recovery procedure.

The magnitude of the entries of v is determined by summing the entries on the diagonal that correspond to the real and imaginary part of the same bus, and taking the square root. After this, the angle difference between buses is calculated based on the filled entries in $\tilde{W}_t^f$. Since the sparsity pattern coincides with the network topology, the filled entries will correspond to the lines of the network. The convex program in Algorithm 7 of extracts the globally optimal voltage vector if $\tilde{W}_t^f$ is rank-one such that $\sum_{(l,m)\in\mathcal{L}} |\angle\tilde{W}_{lm} - \angle v_l + \angle v_m| = 0$. If this is not the case, the program aims to find a voltage vector for which the corresponding angle differences are as close to those suggested by the matrix $\tilde{W}_t^f$ as in [13]. The determined solution is used to build x, which is then used to form W_t^f.

Algorithm 7 Matrix Completion

1: Given: partially filled state matrix $\tilde{W} \in \mathbb{S}^{2N_b}$
2: Initialize: $v \in \mathbb{C}^{N_b}$
3: **for** $k \in \mathcal{N}$ **do**
4: $\quad |v_k| \leftarrow \sqrt{\tilde{W}(k,k) + \tilde{W}(k+N_b, k+N_b)}$
5: **end for**
6: **for** $(l,m) \in \mathcal{L}$ **do**
7: $\quad \angle \tilde{W}_{lm} := \tan^{-1}\left(\frac{\tilde{W}(l+N_b,m) - \tilde{W}(l,m+N_b)}{\tilde{W}(l,m) + \tilde{W}(l+N_b, m+N_b)} \right)$
8: **end for**
9: $\angle v \leftarrow \underset{-\pi \leq \angle v \leq \pi}{\arg\min} \sum_{(l,m)\in\mathcal{L}} |\angle\tilde{W}_{lm} - \angle v_l + \angle v_m|$
10: $x \leftarrow \left[\left(|v| \cos \angle v \right)^{\top}, \left(|v| \sin \angle v \right)^{\top} \right]^{\top}$
11: Output: $W \leftarrow xx^{\top}$

Remark 10.4 It is worth mentioning that the proposed approach in [13] first completes $\tilde{W}$ and then extracts the optimal voltage vector from this completed matrix. We however skip the completion step compared to [13], since our proposed formulation allows us to directly use $\tilde{W}_t^f$ to extract a voltage vector and then reconstruct a completely filled state matrix.

Remark 10.5 Comparing the computational time complexity of CD-OPF-RS with C-OPF-RS by considering that $N_C \ll N_b$, one can clearly see the impact of decomposition on the computational complexity. For realistic networks, N_C is still of reasonable dimensions (see [13] for a list of power systems and their corresponding treewidth, i.e. the size of the largest maximal cliques plus one). CD-OPF-RS has K matrix variables of dimension N_C at most, and the worst-case overall dimension of the matrix variable is therefore $K N_C (N_s + 1) T$.

10.4 Numerical Study

Simulation Setup

We carried out a simulation study using the 30-bus IEEE benchmark power system [30] assuming only a single wind-bus infeed at bus 10. We follow the approach of [31] to generate trajectories for the wind power, with a data-set corresponding to the hourly aggregated wind power production of Germany over the period 2006–2011. The load profile is assumed to be known (see [15, Fig. 6.1]) and the nominal load from MATPOWER[6] [32] is multiplied with this profile to get a time-varying load.

Following Theorem 10.1, we fix $\varepsilon = 10^{-2}$, $\beta = 10^{-5}$, and $N_w = 1$ to obtain the number of required wind power samples at each hour $N_s = 541$. We use Matlab

---

[6]MATPOWER is a commercial software for solving power flow problems using successive quadratic programming.

together with YALMIP [33] as an interface and MOSEK [34] as a solver. All optimizations are run on a MacBook Pro with a 2.4 GHz Intel Core i5 processor and 8 GB of RAM.

After obtaining a solution, the scheduled generator power (the generator power based on the forecasted wind trajectory) and the voltage magnitudes are extracted from $\{W_t^f\}$ for all time steps using the following relations $\forall k \in \mathcal{G}, \forall t \in \mathcal{T}$:

$$p_{k,t}^G = \text{Tr}\left(Y_k W_t^f\right) + p_{k,t}^D - p_{k,t}^w, \tag{10.16a}$$

$$q_{k,t}^G = \text{Tr}\left(Y_k^* W_t^f\right) + q_{k,t}^D, \tag{10.16b}$$

$$|v_{k,t}| = \sqrt{W_t^f(k,k) + W_t^f(N_b + k, N_b + k)}. \tag{10.16c}$$

A comparison using the DC model of power network to solve the OPF-RS problem is delivered as a benchmark approach. A detailed description of DC model can be obtained from [3, 4]. The solution of the benchmark program is the real generator power and distribution vectors for every hour, $\left\{ p_t^{G,\text{dc}}, d_t^{\text{us,dc}}, d_t^{\text{ds,dc}} \right\}$. One also needs the reactive generator power and generator voltage magnitudes in order to have a more realistic comparison. In [8], the nominal value of such variables was extracted from the MATPOWER test case for all time steps and scenarios. This will result in large violations, since the reactive generator power is not adapted to the time-varying demand.

We here develop a novel benchmark approach, namely converted DC (CDC), to have a more sophisticated comparison by solving the following program:

$$\min_{\{W_t\}_{t \in \mathcal{T}}} \sum_{t \in \mathcal{T}} \sum_{k \in \mathcal{G}} \left(p_{k,t}^{G,\text{dc}} - \left(\text{Tr}\left(Y_k W_t\right) + p_{k,t}^D - p_{k,t}^{w,f} \right) \right)^2$$

$$\text{s.t.} \quad W_t \in \mathcal{W}(p_t^{w,f}, s_t^D), \qquad \forall t \in \mathcal{T},$$

$$W_t \succeq 0, \qquad \forall t \in \mathcal{T}.$$

The solution to this program is a feasible (AC) network state $\{W_t\}_{t \in \mathcal{T}}$ where the real generator power is as close as possible to the obtained generator power from the DC solution. The distribution vectors used in simulation will be equal to those obtained from the original solution of the DC framework. A schematic overview of the optimization and simulation process to obtain and validate both the benchmark and proposed formulations is given in Figs. 10.1 and 10.2.

After retrieving a solution, we simulate the network power flow using MATPOWER such that the power and voltage magnitude of generators and all the loads are fixed without imposing any constraints for 10,000 different wind power scenarios. The wind power is implemented as a negative load on the wind-bus. Afterward, the resulting power flows and voltage magnitudes are evaluated by means of counting the number of violated constraints.

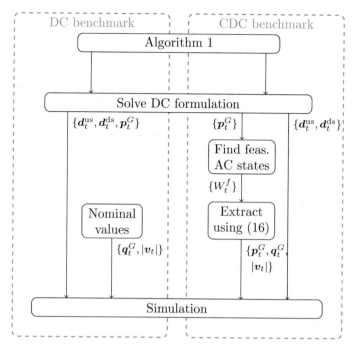

Fig. 10.1 Schematic overview of optimization and simulation process for the DC benchmarks

Fig. 10.2 Schematic overview of optimization and simulation process for the proposed formulations

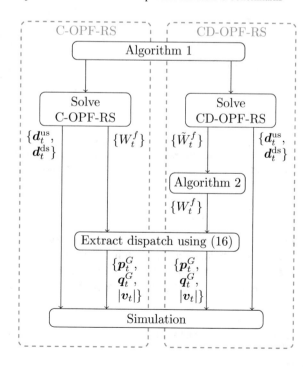

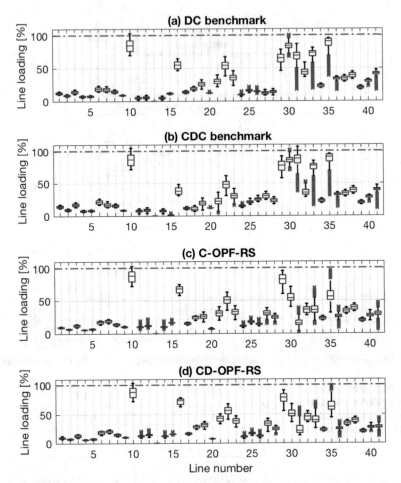

Fig. 10.3 Relative line loading for all hours and scenarios per line. The red line represents the median value, edges of each box correspond to the 25th and 75th percentiles, the whiskers extend to 99% coverage, and the red marks denote the data outliers. The upper plots **a** and **b** show the Benchmark results, and the lower plots **c** and **d** show the proposed approaches

Simulation Results

Figure 10.3 depicts the line loadings[7] as boxplots for DC, CDC, C-OPF-RS, and CD-OPF-RS formulations, for all hours and scenarios. Such a boxplot has been also used in [9] to show line flow violations. It can be seen that all formulations have some violations of the lineflow limits for line 10. To further assess the performance

---

[7]Line loadings are defined as the apparent power flow over a line divided by the maximum apparent power flow for that line, such that a line loading higher than 100% corresponds to the violation of the lineflow limit.

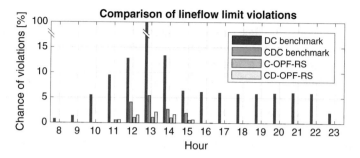

Fig. 10.4 Empirical violation level of lineflow limit for different formulations

of these results, we calculate the level of violation for lineflow limit empirically using the different formulations at each hour (see Fig. 10.4).

In Fig. 10.4 the results for the peak hours are shown. For all other hours, the chance of constraint violation is very close to zero for all formulations. As expected, the DC solution shows a very high level of violation during the peak hours. Although the CDC solution improves the chance of lineflow limit violation, the theoretical limit at the peak hours is still not respected. It is important to notice that the empirical chance of constraint violation for the C-OPF-RS and CD-OPF-RS results are well below the theoretical limit (5%), and they are at most 1.1 and 2.2%, respectively. The proposed decomposition and reconstruction process caused the solution to be slightly less conservative compared to the results of C-OPF-RS.

We next examine the bus voltage magnitudes for all formulations. It is observed that the DC, C-OPF-RS and CD-OPF-RS solutions are always within the limits for all hours and scenarios. However, using the CDC formulation the bus voltage limits show a violation of 100% for all hours. This can be explained by the fact that in the DC framework, the bus voltages are assumed to be constant at nominal value. When we implement the obtained solution in the AC framework, it can be seen that this assumption does not hold. We can thus conclude that for both the DC formulations, the empirical chance of constraint violation is well above the theoretical limits once the solution is implemented in the AC power flow simulations. For both C-OPF-RS and CD-OPF-RS, the a-priori probabilistic guarantees are confirmed to be valid.

10.5 Conclusions

We developed a framework to solve the reserve scheduling (RS) problem using AC optimal power flow (AC OPF) formulation. We first integrated the effect of wind power generation in power networks into an AC OPF problem formulation. Using this new formulation, we unified the RS problem with the AC OPF problem. The final optimization problem leads to an uncertain infinite semi-definite program (SDP) formulation, since the uncertainty set is unknown and unbounded. We approximated the uncertainty set with a-priori probabilistic guarantee using a set-based characterization technique, and then solved a robust finite SDP problem at each hour. A decomposition technique is employed to reduce computational time complexity of the resulting problem. As a final contribution, we proposed a new recovery algorithm to determine a rank-one solution from the decomposed problem. The resulting solutions are validated using Monte Carlo simulations and a commercial power flow simulator, and found to perform as expected. The obtained solutions via our proposed formulation perform better than the solution obtained from the DC framework, which is currently used in industry. Future work will focus on extending the current results to multi-area power systems.

References

1. A. Papavasiliou, S. Oren, R. O'Neill, Reserve requirements for wind power integration: a scenario-based stochastic programming framework. IEEE Trans. Power Syst. **26**(4), 2197–2206 (2011)
2. J. Warrington, P.J. Goulart, S. Mariéthoz, M. Morari, Robust reserve operation in power systems using affine policies, in *Conference on Decision and Control* (IEEE, 2012), pp. 1111–1117
3. K. Margellos, V. Rostampour, M. Vrakopoulou, M. Prandini, G. Andersson, J. Lygeros, Stochastic unit commitment and reserve scheduling: a tractable formulation with probabilistic certificates, in *European Control Conference* (IEEE, 2013) pp. 2513–2518
4. M. Vrakopoulou, K. Margellos, J. Lygeros, G. Andersson, A probabilistic framework for reserve scheduling and N-1 security assessment of systems with high wind power penetration. IEEE Trans. Power Syst. **28**(4), 3885–3896 (2013)
5. G. Andersson, *Modelling and Analysis of Electric Power Systems* (ETH Zürich University, Switzerland, 2008)
6. B. Stott, J. Jardim, O. Alsaç, DC power flow revisited. IEEE Trans. Power Syst. **24**(3), 1290–1300 (2009)
7. J. Lavaei, S.H. Low, Zero duality gap in optimal power flow problem. IEEE Trans. Power Syst. **27**(1), 92–107 (2012)
8. V. Rostampour, K. Margellos, M. Vrakopoulou, M. Prandini, G. Andersson, J. Lygeros, Reserve requirements in AC power systems with uncertain generation, in *Innovative Smart Grid Technologies Europe* (IEEE, 2013), pp. 1–5
9. M. Chamanbaz, F. Dabbene, C. Lagoa, AC optimal power flow in the presence of renewable sources and uncertain loads (2017), arXiv:1702.02967
10. A. Venzke, L. Halilbasic, U. Markovic, G. Hug, S. Chatzivasileiadis, Convex relaxations of chance constrained AC optimal power flow (2017), arXiv:1702.08372
11. A.Y. Lam, B. Zhang, N.T. David, Distributed algorithms for optimal power flow problem, in *Conference on Decision and Control* (IEEE, 2012) pp. 430–437
12. D.K. Molzahn, J.T. Holzer, B.C. Lesieutre, C.L. DeMarco, Implementation of a large-scale optimal power flow solver based on semidefinite programming. IEEE Trans. Power Syst. **28**(4), 3987–3998 (2013)
13. R. Madani, M. Ashraphijuo, J. Lavaei, Promises of conic relaxation for contingency-constrained optimal power flow problem. IEEE Trans. Power Syst. **31**(2), 1297–1307 (2016)
14. V. Rostampour, O. ter Haar, T. Keviczky, Tractable reserve scheduling in AC power systems with uncertain wind power generation, in *Conference on Decision and Control (CDC), 2017* (IEEE, 2017), pp. 2647–2654
15. O. ter Haar, Tractable reserve scheduling formulations for alternating current power grids with uncertain generation. M.Sc. Dissertation Delft University of Technology, The Netherlands, 2017

16. J.M. Morales, A.J. Conejo, J. Pérez-Ruiz, Economic valuation of reserves in power systems with high penetration of wind power. IEEE Trans. Power Syst. **24**(2), 900–910 (2009)
17. S. Boyd, L. Vandenberghe, *Convex Optimization* (Cambridge University Press, Cambridge, 2004)
18. K. Lehmann, A. Grastien, P. Van Hentenryck, AC-feasibility on tree networks is NP-hard. IEEE Trans. Power Syst. **31**(1), 798–801 (2016)
19. Z.-Q. Luo, W.-K. Ma, A.M.-C. So, Y. Ye, S. Zhang, Semidefinite relaxation of quadratic optimization problems. IEEE Signal Process. Mag. **27**(3), 20–34 (2010)
20. R. Madani, S. Sojoudi, J. Lavaei, Convex relaxation for optimal power flow problem: mesh networks. IEEE Trans. Power Syst. **30**(1), 199–211 (2015)
21. G.C. Calafiore, M.C. Campi, The scenario approach to robust control design. IEEE Trans. Autom. Control **51**(5), 742–753 (2006)
22. V. Rostampour, T. Keviczky, Probabilistic energy management for building climate comfort in smart thermal grids with seasonal storage systems, in *International Federation of Automatic Control (IFAC) World Congress* (2017), https://arxiv.org/pdf/1611.03206.pdf
23. V. Rostampour, T. Keviczky, Robust randomized model predictive control for energy balance in smart thermal grids, in *European Control Conference (ECC)* (IEEE, 2016), pp. 1201–1208
24. K. Margellos, P. Goulart, J. Lygeros, On the road between robust optimization and the scenario approach for chance constrained optimization problems. Trans. Autom. Control **59**(8), 2258–2263 (2014)
25. M.C. Campi, S. Garatti, The exact feasibility of randomized solutions of uncertain convex programs. SIAM J. Optim. **19**(3), 1211–1230 (2008)
26. D. Bertsimas, M. Sim, Tractable approximations to robust conic optimization problems. Math. Program. **107**(1–2), 5–36 (2006)
27. R. Grone, C.R. Johnson, E.M. Sá, H. Wolkowicz, Positive definite completions of partial hermitian matrices. Linear Algebr. Appl. **58**, 109–124 (1984)
28. L. Vandenberghe, M.S. Andersen et al., Chordal graphs and semidefinite optimization. Found. Trends® Optim. **1**(4), 241–433 (2015)
29. R. Madani, M. Ashraphijuo, J. Lavaei, SDP solver of optimal power flow users manual (2014)
30. R. Christie, Power systems test case archive, University of Washington (2000), http://www2.ee.washington.edu/research/pstca
31. G. Papaefthymiou, B. Klockl, MCMC for wind power simulation. IEEE Trans. Energy Convers. **23**(1), 234–240 (2008)
32. R.D. Zimmerman, C.E. Murillo-Sánchez, R.J. Thomas, Matpower: steady-state operations, planning, and analysis tools for power systems research and education. IEEE Trans. Power Syst. **26**(1), 12–19 (2011)
33. J. Löfberg, Yalmip: a toolbox for modeling and optimization in Matlab, in *International Symposium on Computer Aided Control Systems Design* (IEEE, 2004), pp. 284–289
34. E. Anderson, The MOSEK optimization toolbox for Matlab manual

Printed in the United States
By Bookmasters